In his long career as a writer Daniel Defoe never tired of advocating the value of personal observation and experience; and he never wavered in his conviction that it is man's God-given duty to explore and make productive use of nature. In this first major study of Bacon's legacy to Defoe Ilse Vickers shows that the ideas and concepts of Baconian science were a significant influence on Defoe's way of thinking and writing. She outlines the seventeenth-century intellectual milieu, and discusses the prominence of Defoe's teacher Charles Morton among major Baconian thinkers of the century. She goes on to consider a wide range of Defoe's work, from the point of view of his familiarity with the ideals of experimental philosophy, and throws new light on the close link between his factual and his fictional works. In the process Vickers reveals a new Defoe not only a thorough Baconian, but also a far more consistent writer than has hitherto been recognised.

CAMBRIDGE STUDIES IN EIGHTEENTH-CENTURY ENGLISH
LITERATURE AND THOUGHT 32

Defoe and the New Sciences

CAMBRIDGE STUDIES IN EIGHTEENTH-CENTURY
ENGLISH LITERATURE AND THOUGHT

General editors

Professor HOWARD ERSKINE-HILL LITT.D., FBA, *Pembroke College, Cambridge*

Professor JOHN RICHETTI, *University of Pennsylvania*

Editorial board

Morris Brownell, *University of Nevada*
Leopold Damrosch, *Harvard University*
Isobel Grundy, *University of Alberta*
J. Paul Hunter, *University of Chicago*
Lawrence Lipking, *Northwestern University*
Harold Love, *Monash University*
Claude Rawson, *Yale University*
Pat Rogers, *University of South Florida*
James Sambrook, *University of Southampton*

Some recent titles

Crime and Defoe: A New Kind of Writing
by Lincoln B. Faller

Locke, Literary Criticism, and Philosophy
by William Walker

The English Fable: Aesop and Literary Culture, 1651–1740
by Jayne Elizabeth Lewis

Mania and Literary Style
The Rhetoric of Enthusiasm from the Ranters to Christopher Smart
by Clement Hawes

Landscape, Liberty and Authority
Poetry, Criticism and Politics from Thomson to Wordsworth
by Tim Fulford

Philosophical Dialogue in the British Enlightenment
Theology, Aesthetics, and the Novel
by Michael B. Prince

Defoe and the New Sciences

ILSE VICKERS

Published by the Press Syndicate of the University of Cambridge
The Pitt Building, Trumpington Street, Cambridge CB2 1RP
40 West 20th Street, New York, NY 10011–4211, USA
10 Stamford Road, Oakleigh, Melbourne 3166, Australia

© Cambridge University Press 1996

First published 1996

Printed in Great Britain at the University Press, Cambridge

A catalogue record for this book is available from the British Library

Library of Congress cataloguing in publication data

Vickers, Ilse.
Defoe and the new sciences / Ilse Vickers.
p. cm. – (Cambridge studies in eighteenth-century English literature and thought; 32)
Includes bibliographical references.
ISBN 0 521 40279 4 (hardback)
1. Defoe, Daniel, 1661?–1731 – Knowledge – Science. 2. Literature and science – England – History – 18th century. 3. Science – Great Britain – History – 17th century. 4. Morton, Charles, 1627–1698 – Influence. 5. Bacon, Francis, 1561–1626 – Influence. I. Title. II. Series.
PR3408.S35V53   1996
823'.5–dc20   96-6229   CIP

ISBN 0 521 40279 4 hardback

For
Beatrice and Gwendolen

Upon the whole, the study of science is the original of learning; the word imports it. 'Tis the search after knowledge.

(Defoe, *The Compleat English Gentleman*)

Contents

List of illustrations	*page* xi
Acknowledgements	xiii
List of abbreviations	xiv

Introduction		1

Part 1 The Baconian scientific milieu

1	The legacy of Francis Bacon	9
2	The selective taking-up of Bacon's ideas; biographical sketches of five followers of Bacon	18
3	Charles Morton and the New Sciences	32

Part 2 Daniel Defoe

4	Daniel Defoe and the Baconian legacy	55
5	Defoe's *General History of Trade*: its relation to the Baconian histories	81
6	*Robinson Crusoe*: man's progressive dominion over nature	99
7	*A New Voyage Round the World*: Defoe the traveller-scientist by sea	132
8	Defoe's *Tour*: a natural history of man and his activities	151

Appendix 177

Bibliography 182
Index 193

Illustrations

1 A diagram illustrating Defoe's 'Nature wrought and altered' in Defoe's *A General History of Trade* (1713). (Courtesy of University of London Library) *page* 90
2 A diagram from Robert Hooke's 'A Scheme representing at one View to the Eye the Observations of the Weather for a Month' from *The Philosophical Transactions of the Royal Society* (repr. 1963). (By permission of the Syndics of Cambridge University Library) 109
3 Edmond Halley's *Magnetic Chart* (1700 repr. 1870). (By permission of the Syndics of Cambridge University Library) 140

Acknowledgements

I would like to thank Philip Oswald for checking the manuscript and spotting inconsistencies and errors. For his careful reading of an early draft of the book and his helpful suggestions, my thanks are due to Maximilian Novak. I shall always be grateful to Brian Vickers whose insightful comments have contributed much to the completion and enjoyment of this project. However, my greatest debt is to Pat Rogers who helped and encouraged me from the start of the undertaking. There is one other obligation that I would like to mention: I owe a special debt to Bill Greenwell whose unfailing friendship has supported me throughout.

In terms of providing opportunities and financial support for research over several years, I am grateful to the Open University, the British Academy and the William Andrews Clark Memorial Library, University of California, Los Angeles.

Finally, I would like to thank my former employer in Brussels, COMETT, the European Community Programme for Education and Training in Advanced Technologies, and particularly my present employer UCLi, University College London *initiatives*, for so generously giving me study leave to complete this work.

The book is dedicated to my daughters Beatrice and Gwendolen whose love has contributed to every page.

For any errors and oversights I am solely responsible.

Abbreviations

Works by Defoe

Works by Defoe referred to more than once are given in the following abbreviated form. The place of publication is London, unless otherwise stated.

Atlas Maritimus and Commercialis   *Atlas Maritimus and Commercialis; or, a General View of the World, so far as it relates to Trade and Navigation* (1728)
Augusta Triumphans   *Augusta Triumphans: or, the Way to make London the most flourishing City in the Universe* (1728)
Brief State   *A Brief State of the Inland or Home Trade of England* (1730)
Caledonia   *Caledonia, a Poem in Honour of Scotland, and the Scots Nation* (Edinburgh, 1706)
The Consolidator   *The Consolidator: or, Memoirs of Sundry Transactions from the World in the Moon* (1705)
Crusoe 1 and 2   *The Life and Strange Surprizing Adventures of Robinson Crusoe, of York, Mariner* (1719)
Crusoe 3   *Serious Reflections during the Life and Surprising Adventures of Robinson Crusoe* (1720)
Gentleman   *The Compleat English Gentleman* (ed. Karl D. Bülbring, 1890)
Great Law   *The Great Law of Subordination consider'd; or, the Insolence and Unsufferable Behaviour of Servants in England duly enquir'd into* (1724)

Historical Account of Sir Walter Raleigh   *An Historical Account of the Voyages and Adventures of Sir Walter Raleigh* (1720)

History of Arts and Sciences   *A General History of Discoveries and Improvements, in useful Arts, particularly in the great Branches of Commerce, Navigation, and Plantation, in all Parts of the known World* (1725–7)

History of Peter Alexowitz   *An Impartial History of the Life and Actions of Peter Alexowitz, the Present Czar of Muscovy* (1723)

History of Trade   *A General History of Trade* (1713)

Humble Proposal   *An Humble Proposal to the People of England, for the Encrease of their Trade, and Encouragement of their Manufactures* (1729)

Mere Nature Delineated   *Mere Nature Delineated: or, a Body without a Soul* (1726)

More Short-Ways   *More Short-Ways with the Dissenters* (1704)

A New Voyage   *A New Voyage Round the World by a Course never sailed before* ('1725' for 1724)

Plan of Commerce   *A Plan of the English Commerce* (1728)

Present State   *The Present State of the Parties in Great Britain: Particularly an Enquiry into the State of the Dissenters in England, and the Presbyterians in Scotland*... (1712)

Projects   *An Essay upon Projects* (1697)

Review   *A Weekly Review of the Affairs of France and various similar titles*, 9 vols. (1704–13)

Shortest-Way   *The Shortest-Way with the Dissenters: or Proposals for the Establishment of the Church* (1702)

The Storm   *The Storm: or, a Collection of the most remarkable Casualties and Disasters which happen'd in the Late Dreadful Tempest, both by Sea and Land* (1704)

Tour   *A Tour thro' the Whole Island of Great Britain* (1724–6)

Tradesman   *The Complete English Tradesman*, 2 vols. (1726–7)

True-Born Englishman   *The True-Born Englishman. A Satyr* (1701)

Other references

DNB	*Dictionary of National Biography*
DSB	*Dictionary of Scientific Biography*
EIC	*Essays in Criticism*
ELH	*A Journal of English Literary History*

HLQ	*Huntingdon Library Quarterly*
JHI	*Journal of the Histories of Ideas*
MP	*Modern Philology*
OED	*Oxford English Dictionary*
PQ	*Philological Quarterly*
RES	*Review of English Studies*
SP	*Studies in Philology*

Books and articles are listed in full in the bibliography and are referred to in the text in abbreviated form, giving the author's surname and date as follows: Earle 1976: 43.

Introduction

Defoe is regularly discussed as the 'Father of the English novel', a pioneer in modern economic thought, England's first journalist, and in other similar terms which emphasise his influence on succeeding ages down to the present. Few scholars remember, so it seems, that he was born in the middle of the seventeenth century and had lived for almost forty years (more than half his lifetime) before the appearance of his first major publication, *An Essay upon Projects* (1697), and the first issue of the *Review* in 1704. This neglect of his seventeenth-century roots is the more regrettable since Defoe himself frequently referred to the time of his youth and education, was demonstrably backward-looking in his economic ideas, and, furthermore, chose to set the majority of his fictional works in the period from 1660 to 1680. While the knowledge of what Defoe inspired in others has increased our understanding of his work, it has tended to obscure one entire dimension of his work. This book tries to answer this second need. It is intended to complement, not to deny the value of the research done so far. By looking at what the author inherited I hope to add something to our appreciation of Defoe as an educational and economic commentator and creative writer.

Essentially this book is an attempt to outline the seventeenth-century Baconian tradition and to define Defoe's place in this movement of ideas. To be sure, by the time Defoe published his first novel in 1719 many traditions had contributed to shaping his mind. While concentrating on the Baconian influence, I do not claim that it was the sole influence, but merely one which has been unjustifiably neglected.

It was at the famous Academy for Dissenters run by Charles Morton that Defoe first came into contact with the principles of Baconian experimental science. In order to re-create the historical situation that confronted Defoe at Morton's Academy, I take Francis Bacon as my starting-point. In a brief outline of the most influential elements of Bacon's philosophy I show that his instauration, or renewal, of knowledge was closely bound up with educational, social, economic and linguistic programmes of reform in the period from 1640 to 1670. The following accounts of the thought of five prominent Baconians – Samuel Hartlib, William Petty, Robert Boyle, Robert Hooke and John Wilkins – demonstrate that different as these scientists were they shared Bacon's ideals of direct observation of nature and man, systematic recording of data, and a belief in the value of science for the 'relief of man's estate'. Turning from the first-generation Baconians to Charles Morton, we can see in more detail than has yet been provided that Defoe's teacher was intimately acquainted with the research of the experimental scientists of his day, namely Boyle, Petty, Hooke and others. In addition, Morton went back directly to Bacon's philosophy, and it is through this source that Defoe himself became a committed Baconian. In this context I also discuss the issue of the proper language for science, showing that Morton's recommendation of 'plain' prose was derived from Bacon and from a work in the Baconian tradition, *Ecclesiastes; or, A Discourse concerning the Gift of Preaching, as it falls under the Rules of Art* (London, 1646) by John Wilkins.

The second part of the work is devoted to Defoe himself, beginning with a general account of his debt to Bacon. Perhaps no other aspect of Defoe's writing gives us as clear an insight into his habit of mind as his practical approach to learning. His views on education are partly autobiographical, recalling the teaching at Morton's Academy, and partly derived from the Baconian Puritans (such as Hartlib and John Dury). Common to this whole tradition, culminating in Defoe, was the belief that knowledge should be useful to society, and be concerned with 'things, rather than words'. Works in which Defoe's use of Baconian aims and methods is especially visible are *The Storm* (1704);

The Consolidator: or, Memoirs of Sundry Transactions from the World in the Moon (1705); and *A General History of Discoveries and Improvements, in useful Arts* (1725–7). The first of these is an account of the famous tempest in 1703, made up of direct reports and published descriptions from the *Philosophical Transactions of the Royal Society*, which moves from factual recording to the establishment of a theory concerning the origin of storms. The second work, apparently a lunar fantasy, in fact demonstrates Defoe's wide knowledge of the activities of the Royal Society. The third work is a typical Baconian 'History', that is, a systematic inquiry into a field of knowledge with suggestions for reform.

Defoe's fullest essay in this genre is his *General History of Trade* (1713). This work fulfils the purpose Bacon called for in the 'History of Nature Wrought or Mechanical', namely a 'History of Trades' or survey of all human activities, from arts to craft techniques and technologies. This was one of Bacon's most influential proposals and many of his followers such as Boyle, Evelyn, Hooke and Petty started to compile such histories. Defoe's contribution shifts the focus from trades, in the sense of crafts or vocations, to trade as a commercial activity, ideally involving the whole nation in an exchange process from which all would benefit. Defoe emerges here as an advocate of social reform along Baconian lines.

Since Defoe did not believe in systematically ordering his views under specific headings, it is not always easy to discover whether and how his line of thought changed over the years. Studying a range of his writings from between 1697, the year of his first important published work, and his death in 1731, and collecting his statements on different subjects, such as trade, education, social and economic reform, it becomes quickly evident that many of Defoe's ideas were formulated early on in his life. On many central issues Defoe defended the same view over a period of more than thirty years.

Usually the author's output is categorised into factual and fictional works. Exploring his link with seventeenth-century science, it has become clear that Defoe himself was unaware of such a division. Furthermore, there can be little doubt that his polemical works motivated or inspired him to write fiction. It is in Defoe's non-fictional tracts, heavily Baconian in character, that we have our most important source for our understanding of his fictional works. For it is here, especially in the works on trade written just before, during and after Defoe's enormous output of fiction, that we are best prepared for Crusoe's, Moll's or Roxana's view of the world. That Defoe was influenced by Bacon's new philosophy has not gone unnoticed.[1] But a systematic investigation that considers his alignment with the main Baconian principles and his translation of these ideas into fiction has not been thought worthwhile.

Various critics have commented on such practical aspects of Crusoe's activities as the tools and instruments which enable him to reassert his dominion over nature. More important, Defoe endows his hero with the Baconian mentality of observation and experiment, learning step-by-step from his mistakes the whole process required to bake bread, to make baskets, pots, clothes, and other artefacts. Crusoe keeps a Baconian table of the weather, observes the tides, and maintains a diary of all remarkable events – as recommended by the Royal Society. Crusoe works not just as a scientist but as a Christian, sharing the belief of many seventeenth-century scientists that the study of the created world would give insight into the divine order. An important aspect of *The Life and Strange Surprizing Adventures of Robinson Crusoe, Of York, Mariner* (1719) is the relation between making and writing. In Crusoe's written narrative, constantly recurring words are 'make' and 'thing(s)'. Language, too, is made by the most direct means, putting words to the service of things, following the Baconian principles for scientific prose.

These Baconian ideals also give the appropriate form to the last two works studied here, *A New Voyage round the World* ('1725' for 1724) and *A Tour thro' the Whole Island of Great Britain* (1724–6). Travel had from the

---

[1] For example, John McVeagh acknowledges that Defoe 'rarely mentions Bacon. But he is Bacon's true heir.' Pointing out the relationship between the *Historical Account of Sir Walter Raleigh*, *Crusoe* 1 and *Crusoe* 2, he writes that 'it is likely that this theme, the Baconian theme, the possibility of improving men's condition by the processes of discovery, agriculture and trade, was in all of them Defoe's primary reason for writing; at any rate, through them all the vision is held' (McVeagh 1981: 55–7). Barbara Shapiro, studying the link between the Baconian natural history, the travel book and fiction at the beginning of the eighteenth century, observes that 'Scholars have not yet fully disentangled the invented and noninvented materials' in Defoe's work. She sees Defoe's imaginative work as merging 'imperceptibly with his *A Tour Through the Whole Island of Britain*, a travel book full of precise descriptions of natural resources and economic and social life. This volume ... easily fits into the category of the travel book and natural history' (Shapiro 1983: 263). And see Baridon 1984, and Backscheider 1985. For my article relating the Baconian histories to Defoe's *General History of Trade* and *Robinson Crusoe* see I. Vickers 1987: 200–18.

start been an important aspect of the new philosophy : it was one of the means whereby Bacon's demand for personal observation and collection of data could be realised. In Defoe's travel books we have perhaps the author's most practical applications of the fundamental tenets of the New Sciences. In part two of *A New Voyage* Defoe uses and adapts the Royal Society's directions for travellers by sea to write his propaganda piece for colonisation in South America. The *Tour* is a topographic 'History', an itinerary which records the most noteworthy things about counties, towns, ports – the population, trade, prosperity or lack of it, the characteristics of the inhabitants. In effect the *Tour* is a natural history of man and his activities. While presented as first-hand observations, it has long been known that Defoe relied on printed sources, especially Camden's *Britannia*, in the enlarged edition of Edmund Gibson (London, 1695). Although Defoe openly acknowledges Camden/Gibson as his source on some eighty occasions, there are many more silent borrowings, particularly of the 'Additions' supplied for the 1695 edition. The significant fact about these additional sections is that many of the authors were Fellows of the Royal Society, and represent the continuing influence of the Baconian tradition with its care for accurate observation. Although this discovery diminishes Defoe's claims to originality, it shows his identity with some of the main emphases in seventeenth-century science.

The texts that I have used are listed in the bibliography. For the canon I have generally relied on J. R. Moore's *Checklist*, bearing in mind that many attributions are still controversial. I have also benefited from P. N. Furbank's and W. R. Owens's *The Canonisation of Daniel Defoe* (London, 1988).

I occasionally refer to the Sales Catalogue of Defoe's library – or rather, the joint catalogue recording the sale on 15 November 1731 by the book-seller Olive Payne of two libraries, that of Defoe and that belonging to Phillips Farewell (1688–1730). Farewell was a clergyman, Fellow of Trinity College, Cambridge, and it is interesting to speculate on which books belonged to which collector. Of more than 2,000 titles many are on divinity and ecclesiastical affairs, law, history, political affairs and theory. It is particularly tantalising for my interests to note that many volumes are on the New Sciences. Physics, chemistry, botany, natural history, zoology are represented with approximately 60 works, while the books on industries, crafts, gardening, husbandry, trade and commerce amount to about 100. Titles relating to geography, topography, maps and travels amount to more than 150 and there are 85 books on medicine. My impression is that these works are more likely to have belonged to Defoe, but one cannot be certain, and for that reason I have not based my argument on them. In his modern edition, Helmut Heidenreich has shown in some detail the coherence between many of the books in the collection and Defoe's own interests (Heidenreich 1970: xvii–xxviii). In the Appendix I list some of the titles especially relevant to my argument.

PART 1

The Baconian scientific milieu

1

The legacy of Francis Bacon

I

'For Defoe, above almost any other, we may recall the words of Sir Francis Bacon: "he is a citizen of the world, and ... his heart is no island cut off from other lands, but a continent that joins to them".' J. R. Moore's biography of Daniel Defoe concluded with this sentence; I shall make it my starting-point.[1]

In the seventeenth century Bacon (1561–1626) was revered as the 'mighty Man', who, like Moses, had led his age out of the 'barren Wilderness' of antiquity to 'the very Border' of a new age.[2] As modern studies have shown, Bacon was not a fully-fledged scientist but a propagandist for the reform of scientific thought and method. Realising that traditional methods of investigation failed to secure intellectual progress, he put forward a new plan for the instauration (or renewal) of science which, he demanded, 'must be laid in natural history ... of a new kind and gathered on a new principle'. These ideas were first published in *The Advancement of Learning* (London, 1603); they were repeated in the *Instauratio Magna* (London, 1620), containing his *Novum*

---

[1] From Bacon's *Essays*, 'Of Goodness, and Goodness of Nature', Bacon 1857, VI: 405; Moore 1958: 334.
[2] From Abraham Cowley's 'Ode to the Royal Society', prefixed to Sprat's *History of the Royal Society* (London, 1667). On Bacon's influence in the seventeenth century see Jones 1961, C. Webster 1975, Hill 1965 and M. Hunter 1981.

Organum or 'New Instrument' of scientific method. Although some of Bacon's major ideas had been expressed before, he was the first to collect and incorporate them into an effective programme for the reform of learning.

Bacon's plan for the reformation of knowledge began by exposing the inadequacies inherent in earlier 'philological' science. He drew attention not only to the uselessness but the evil of believing that truth was only to be found in written authority of the greatest classical philosopher, Aristotle. 'Such teachings,' Bacon argued, 'if they be justly appraised, will be found to tend to nothing less than a wicked effort to curtail human power over nature and to produce a deliberate and artificial despair. This despair in its turn confounds the promptings of hope, cuts the springs and sinews of industry, and makes men unwilling to put anything to the hazard of trial.'[3] Intellectual progress was hindered by the purely verbal, syllogistic method of reasoning in Aristotelian logic (as systematised in the Middle Ages), which, Bacon argued, could prove only that which was already known without producing new ideas (IV: 24). Rejecting these limitations, he urged man to begin afresh by studying nature directly and in all its aspects, including those which so far had been regarded as too vulgar or trivial. The new programme, based not on books and established authorities but on careful observation, collection and classification of data gave a new impetus to science; it restored in men the hope and belief in their own ability to make new contributions to scientific progress.

Bacon's high esteem for personal observation and experience may be seen from the advice he gives that experiments should be performed, if possible, by the practitioner himself. If second-hand observations had to be relied on, they should be marked with 'a qualifying note, such as "it is reported"', and the source be given. Experiments should be described with utmost care so that they could be repeated, tested and improved (IV: 260). Equally important for the re-direction of efforts was Bacon's emphasis on an open, sceptical attitude which accepted nothing on trust. According to the new way of thinking, a man must be content to begin in doubts, if he was to end in certainties; his field of investigation must embrace the totality of human experience (IV: 256,

---

[3] III: 592 and 594; I have made use of the translation of B. Farrington 1973: 152.

III: 293). The sceptical or questioning habit of mind guards against making too hasty judgements: 'He who makes too great haste to grasp at certainties shall end in doubts, while he who seasonably restrains his judgment shall end in certainties' (III: 250; IV: 357–8, 429). Laying down general precepts upon which the 'Natural and Experimental History' for a 'true philosophy' was to be built, Bacon explained that in the history which he designed, 'special care is to be taken that it be of wide range and made to the measure of the universe. For the world is not to be narrowed till it will go into the understanding (which has been done hitherto), but the understanding to be expanded and opened till it can take in the image of the world, as it is in fact' (IV: 255–6). Bacon's comprehensive registers of experiments and observations were to provide the material for a new scientific method, rising from specific instances to 'middle axioms' and hence to scientific laws.[4]

It should be emphasised that Bacon used the word 'history' not as a chronologically based narrative but in the available sense (as seen in Aristotle's *Historia Animalium*), that is, as a systematic study of a complete field of knowledge. If we turn to the *OED* we find under the fifth heading the explanation that best describes the Baconian programme: a 'history' is a 'systematic account (without reference to time) of a set of natural phenomena, as those connected with a country, some division of nature or group of natural objects, a species of animals or plants, etc. Now rare except in NATURAL HISTORY.' As a systematic inquiry into all the branches of human knowledge, the Baconian history was meant to be both a repository of what was known, and also to offer 'anticipations' or suggestions for future research.

'The true and lawful goal of the sciences is none other than this: that human life be endowed with new discoveries and powers. But of this the great majority have no feeling, but are merely hireling and professorial; except when it occasionally happens that some workman of acuter wit and covetous of honour applies himself to a new invention; which he mostly does at the expense of his fortunes' (IV: 79). Bacon anticipated that intellectual progress depended on collaboration and co-operation. If, as R. F. Jones rightly pointed out, the most

---

[4] For a concise account of Bacon's scientific method see M. Hesse's 'Francis Bacon's philosophy of science' in B. Vickers 1968: 114–39.

influential element of Bacon's 'philosophy was his conception of a natural history' (Jones 1961: 18), then this was due mainly to two reasons: first, the need to collaborate; second, the realisation that not only those with exceptional talents but everybody, even the 'illiterate' craftsmen, could contribute with their first-hand experience to the renewal of science. Bacon argued that his collaborative enterprise, to be completed over many years and by many hands, would go 'far to level men's wits, and leaves but little to individual excellence; because it performs everything by the surest rules and demonstrations' (IV: 62–3, 109). The work that was to show the full structure of Bacon's intertwining ideas was the *New Atlantis* (written *c*. 1624, published by his executors after his death). In this utopian vision Bacon outlined his ideal of a scientific research institute, where all branches of human knowledge were represented, and members combined to collect data, perform experiments, develop scientific theories, and to share their findings as well as their problems. Bacon gave this institute the name of 'Salomon's House, or the College of the Six Days' Works' – it became the blueprint or 'Prophetick Scheam', as Glanvill put it, according to which the Royal Society of London was founded in the 1660s (Glanvill 1885: lxv).

Nearly everything that Bacon advocated concerning the advancement of learning was actuated by his belief that scientific pursuits should lead to technological progress. In Bacon's own words: 'Man is but the servant and interpreter of nature: what he does and what he knows is only what he has observed of nature's order in fact or in thought; beyond this he knows nothing and can do nothing. For the chain of causes cannot by any force be loosed or broken, nor can nature be commanded except by being obeyed. And so those twin objects, human Knowledge and human Power, do really meet in one' (IV: 32). Bacon laid down that, of all the natural histories, the 'History of Nature Wrought or Mechanical' – that is, a history of arts and crafts as they have been applied to nature – was the most important. Among the areas of research he proposed, we find recommendations for a 'History of Stone-cutting', 'Of the making of Bricks and Tiles', 'Of Pottery', 'Of Cements' and so on (IV: 269–70). Bacon died leaving his grandiose scheme for a history of human activities uncompleted. Apart from his detailed advice to his followers on how to carry out the great collaborative enterprise for the reform of knowledge, he supplied some specimen histories, namely a 'History of the Winds', a 'History of Life and Death' and a 'History of Dense and Rare' (V: 137–200; V: 213–335; V: 337–400). But his main contribution consists in leaving his successors with a clear outline of the methods and goals of the natural history programme.

As is abundantly evident, Bacon did not envisage his natural history as a random amassing of material. The aim of the carefully ordered registers of facts was twofold: the first was concerned with the 'speculative' or with 'Anticipations of the New Philosophy', that is, predictive theories based on the discovery of the fundamental, unalterable laws of nature. The second, applying and testing these tentative generalisations to works and science, represented the 'operative' aspect of Bacon's new method. But the two aspects interact, forming the basis of the last three stages of his *Great Instauration*. The scientist proceeds, as by a 'Ladder of the Intellect': applying the inductive method he moves from observations and experiments to axioms and hypotheses, then re-applying his findings to some practical goals. 'True and fruitful Natural Philosophy has a double scale or ladder, ascendent and descendent, ascending from experiments to axioms and descending from axioms to the invention of new experiments' (IV: 343). Although Bacon considered both branches of his philosophy of importance, his followers did not take up his elaborate scientific method, but preferred to concentrate on the natural 'histories' and on experiments that led to 'use or discovery'.

'Let us hope', Bacon wrote, that 'there may spring helps to man, and a line and race of inventions that may in some degree subdue and overcome the necessities and miseries of humanity' (IV: 27). Whereas the Greeks had valued knowledge for its own sake, and the medieval church discountenanced intellectual inquiry into those areas which could infringe on God's power and mystery, Bacon legitimised knowledge for its power to improve the quality of life. His whole cast of thought was addressed to 'balancing the miseries of man', and he impressed upon his age the importance of 'useful', 'practical' knowledge – expressions which were used over and over again throughout the century following. If Bacon advised men to direct knowledge 'for the benefit and use of life', he equally stressed that it should be 'perfected and governed [by] charity'; knowledge should be 'a rich storehouse, for the glory of the Creator and the relief of man's estate' (III: 294). It is important that we remember that 'Bacon professed no such narrow utilitarianism as later went under his name' (Hill 1965: 93), and that his 'utilitarian' end for science was derived from 'the older traditions of the *vita activa* or involvement in society for the benefit of others (an idea that goes back to Plato and Cicero), and Christian charity' (B. Vickers 1987: 2–3). Drawing on these established traditions, Bacon suggested a method that would not only 'endow the condition and life of man with new powers and works', but claimed to ensure man's 'Empire over things' (IV: 104, 114). Bacon's confident hope that with the method he proposed man could illuminate every aspect of the universe, firmly impressed itself upon English science. Bacon 'did more than anyone else to break the fetters which bound his age to servile submission to the authority of the ancients, and he inspired his followers to face the future rather than the past ... His was a stimulating and vitalizing influence hardly to be over-estimated' (Jones 1961: 60–1).

Associated with Bacon's 'utilitarian' – in the sense of philanthropic – concept of learning was his conviction that it was God's divine plan that man should explore nature and learn to control her to his advantage. Bacon warned that 'if any man shall think by view and inquiry into these sensible and material things, to attain to any light for the revealing of the nature or will of God, he shall dangerously abuse himself' (III: 218). While scientific inquiries into nature could not reveal 'the image of the Maker', science itself was neither a blasphemous nor an unlawful activity. Bacon separated science from religion, and yet he acknowledged that science did not undermine Christian faith. Just the opposite: he put forward the astonishing suggestion that experimental science held within itself the power of repairing the Fall and regaining for man his dominion over things. 'For man by the Fall fell at the same time from his state of innocency and from his dominion over creation. Both of these losses, however, can even in this life be in some part repaired; the former by religion and faith, the latter by arts and sciences' (IV: 247–8). For Bacon the 'study of the book of God's Word' and 'the book of God's works' conjoined, would help to reveal the wisdom of God in his creation. Echoing Bacon, the Christian virtuosi later in the century asserted that 'by being addicted to Experimental Philosophy, a Man is rather assisted than Indisposed, to be a good Christian' (Boyle 1774, V: 37). Bacon's reform programme leads directly to the kind of natural theology that we find in men such as Boyle, Wilkins and Ray.

Bacon frequently referred to the Bible to find justification or parallels for what he was doing. His greatest interest was in King Solomon. On many occasions he declared that 'Salomon the king, as out of a branch of his wisdom extraordinarily petitioned and granted from God, is said to have written a natural history of all that is green from the cedar to the moss ... and also of all that liveth and moveth.' Solomon's conscientious compiling of natural histories was rewarded, for God granted him power over nature and a 'mighty empire of Gold'. Admiring Solomon's search after the knowledge of things, Bacon declared that 'the glory of God *is to conceal a thing, but the glory of the king is to find it out* '.[5] Solomon served Bacon as a pattern for his natural history programme, and he called on his age to emulate the king. Responding to this call, his disciples enthusiastically embraced the task of 'digging in the mines of nature', or, as they variously described it, 'unlocking' the 'treasure-house' of nature, to bring to light new facts of experience.

II

In this book I shall argue in some detail Defoe's indebtedness to the Baconian revolution.[6] However, at the outset it seems appropriate to indicate some general similarities and differences. Looking at Bacon's whole scientific programme, it is clear that Defoe is not equally interested in every aspect of it. Defoe is not interested in induction or aphorisms, nor in Bacon's scientific method as a totality. The aspects of Bacon's philosophy in which he is interested are those which an ordinary man can do and use. Defoe observes and collects data of physical reality and applies this information to investigations of economic and social life. In other words, he uses the observational

---

[5] III: 219–20, see also IV: 114; and cf. 1 Kings 4: 33–4, and Proverbs 25: 2.
[6] Princeton Library houses an annotated copy of the *Advancement of Learning*. Paula Backscheider in a very interesting article suggests that this is in fact Defoe's copy. I have studied the marginalia but for the moment, particularly until we have a graphologist's opinion, would like to reserve my final judgement (Backscheider 1985).

method and directs it to 'the benefit of man's life' but then stops short; he does not, as the Baconian scientist was invited to do, rise from phenomena up the ladder of axioms to natural laws and general principles of natural philosophy.

Defoe re-engages in the warfare of the 'moderns' against the 'ancients' but he is not interested in Bacon's criticism of Aristotelian science, this battle having been fought and won long before Defoe began to write. Defoe's campaign against the ancients is derived rather from the second half of the century when the quarrel had resulted into clear groupings contesting two antithetical concepts of knowledge, culminating in Swift's *Battle of the Books* (London, 1704; cf. Jones 1961, and Levine 'Ancients, moderns and history', in Korshin 1972: 43–75). Although the nature of the battle had changed, Defoe retained many of Bacon's initial arguments dismissing the ancients' method as erroneous and elevating the moderns' as the 'door' or means leading to the promising future. Defoe writes that the ancients have 'led us by the Hand to the very Door, where what remains is to be found'. The past knew not the value of 'Experiment', 'Instruments', 'Scale', 'Demonstration', and so on; they had not tried the new method by which the moderns, 'as if all Nature was newly laid open to them, make daily more and more Discoveries in the Principles of things'. Pronouncing 'experimental as well as naturall phylosophy the most agreeable as well as profitable study in the world', Defoe stands in line with the Baconian programme (*A General History of Discoveries and Improvements* (1725–7), Preface, pp. 233–4, 266; *The Compleat English Gentleman* (published posthumously in 1890), p. 228 and see pp. 65–80 below).

Defoe in effect dedicates himself to one of Bacon's main themes, the production and promotion of knowledge useful to society at large. To this end he compiles compendious registers of facts, that is Baconian 'histories' of nature altered by the activities of man. Explaining the purpose of his *History of Arts and Sciences*, he writes that it is principally intended to 'instruct', 'particularly as it will kindle new desires after the farther Discoveries, and Improvements ... useful for the good of Mankind' (pp. v–vi). Defoe believes in education. His aim is, as he frequently repeats, 'to open people's eyes', and call them to an awareness not only of the need but of the possibility of improving their social and economic situation. His desire is to awaken men from their 'long *Lethargick* Dream' of despondency and ignorance. Both in his role as adviser and in the remedies he proposes to combat the evil, Defoe defends an attitude essential to the New Sciences (see pp. 96–8 and 169–73 below). As we have seen, Bacon had encouraged his age to follow King Solomon in his diligent and systematic search after knowledge. It seems that one of those who at the beginning of the eighteenth century still heard Bacon's call was Defoe. Praising Solomon's unwearied 'resolution of improving himself', Defoe in his turn urges his countrymen to work 'according to Solomon' and '*Search for knowledge as for silver and dig for it as for hid treasure*' (*Gentleman*, pp. 37, 212). So far, Defoe's many references to Solomon, suggesting his understanding of Bacon's experimental science, have gone unnoticed. This is surprising since we find them in the early and in the major works of the maturer years, in Defoe's non-fictional writings as well as in his fiction (see pp. 116–17, and pp. 175–6 below). However, the channels by which Bacon's influence was propagated and reached Defoe is the subject of the following chapters.

2

The selective taking-up of Bacon's ideas; biographical sketches of five followers of Bacon

In the selective taking-up of the Baconian ideas after 1640, two main 'scientific' centres of interest can be discerned. They are concerned with the two distinct aspects of knowledge as Bacon envisaged them: the 'fruit' and 'light' of learning, that is, the production of useful works and new discoveries. On the one hand, there evolved a group of men who, while not being genuine, 'practising' scientists, dedicated their energies to the promotion of ideas connected with science. This circle was associated with Samuel Hartlib and included among others John Dury, John Amos Comenius (in scholarly context) or Komensky, John Evelyn and (to begin with) the young William Petty and Robert Boyle. Developing the Baconian recommendation that knowledge must be shared, Hartlib and his group saw themselves in the first instance as co-ordinators and disseminators of information.

The other group, headed by John Wilkins, followed more obviously in Lord Verulam's footsteps. They were the natural philosophers or scientists, men like Boyle, Petty, Hooke, Wallis, who – initially as the 'Invisible College' and then as the Oxford group – were the leaders of the English scientific revolution.[1] While this group concerned itself with 'experimenting' and the discovery of fundamental scientific laws,

---

[1] Scholars are not agreed whether the 'Invisible College' actually refers to Hartlib's circle: see Syfret 1947–8 and Turnbull 1952–3.

Hartlib's circle became involved in reform-schemes for the amelioration of society.

One can see the two groups as the active, 'scientific' reformers, as opposed to the intellectual and highly specialised 'pure' new scientists. Perhaps a more valuable approach would be to recognise the social and intellectual differences that determined the character of the two settings. The reformers worked at a 'lower', more democratic level (they were schoolmasters, chaplains with Puritan affiliations), while the Oxford group consisted largely, though not exclusively, of sons of gentry; they were research scientists and academics; they were royalists and belonged, by and large, to the Church of England. Yet the distinction between the two groups should not be over-emphasised. Both acknowledged their intellectual debt to Bacon; both criticised the sterile methods of the past and advocated the direct study of reality; both shared the belief in improvement by progressive stages of certainty. It is important to realise that both Hartlib's circle and Wilkins's contributed to the foundation of the Royal Society. Long before the Society was called into being, these groups created through their regular meetings and correspondence a 'rendez-vous' for the discussion of knowledge for the benefit of all.[2]

The following biographical outlines are intended to give an idea of the social and intellectual backgrounds of five prominent Baconians. I have chosen Hartlib, Petty, Boyle, Hooke and Wilkins to show the very different ways in which the new philosophy could be taken up and developed. My choice was further determined by the fact that we have reason to believe that Defoe was acquainted with certain aspects of these Baconian scientists, since they are represented with one or more works in the Sales Catalogue of the Defoe/Farewell libraries.[3] However, the main purpose for giving these biographical sketches is to outline the system of thought which characterises the first-generation Baconians – a system of thought against which Defoe's mind can be measured.

---

[2] For a more detailed analysis of the two groups see Houghton 1941: 39 and C. Webster 1975: 498–500.
[3] See the Appendix pp. 177–81.

Samuel Hartlib (*c.* 1598–1662)

Hartlib came to England from Poland towards the end of the 1620s when he was about thirty years old.[4] Celebrated as 'the Great Intelligencer of Europe', Hartlib soon came to know the most influential people of the day engaged in scientific, educational or ecclesiastical learning. He counted among his friends Milton, Petty, Boyle, Wren, Wilkins, Seth Ward, John Dury, Hobbes, Evelyn, Pepys, and abroad he corresponded with Descartes, Mersenne, Comenius, Huygens, Hevelius, de Greer and many others. Hartlib became a clearing-house for ideas; anyone interested in the New Sciences before 1660 would naturally turn to Hartlib for assistance.

Hartlib's main goal was to co-ordinate and systematise the study of nature. In this effort he encouraged his friends to commit their ideas to paper and he often promoted their works at his own expense. It was Hartlib who asked Milton to compose his treatise *Of Education* (London, 1644); in 1655 Christopher Wren wrote about a transparent beehive at Hartlib's instigation (Hartlib published this in *The Reformed Common-Wealth of Bees* (London, 1655)); whilst Hartlib was also responsible for Boyle's first published work, 'An epistolical discourse of Philaretus to Empiricus', written anonymously in 1647.

Not interested in descriptive natural history, Hartlib and his associates adapted the basic Baconian principles to investigations of husbandry, navigation, trade and education. They studied nature but it was nature 'wrought', altered by the activities of men. In 1648 Hartlib published *A Further Discoverie of the Office of Publick Addressse*, a scientific study of every aspect of the country which – so Hartlib hoped – would yield ideas for economic and social improvements. While there seems no direct link with the natural history of trades, Bacon's advice for such a scheme is right at the centre of Hartlib's work. Encouraged by Hartlib, Petty wrote his *Advice ... for the Advancement of some particular Parts of Learning* (London, 1648) which contained an outline of a History of Trades. Three years later appeared Hartlib's *Legacy of Husbandry* in which he recommended the use of Baconian natural history for

---

[4] The exact date of Hartlib's birth is not known; it is assumed that he was born in the last decade of the sixteenth century. The following biographical outline is chiefly based on Turnbull 1920, Turnbull 1947 and Turnbull 1952–3; Syfret 1947–8, and C. Webster 1970.

agricultural reform. Hartlib's *Legacy* became 'one of the most important agricultural writings of the century' (C. Webster 1975: 473).

Two of Hartlib's associates need to be singled out: John Amos Comenius (1592–1670) and John Dury or Durie (1596–1680). Known as the 'pansophists' (from Comenius' endeavours towards universal knowledge, 'pansophia'), these three worked for reform in all spheres of life. Educational reform was an important aspect of their schemes.[5] Combining Baconian and Puritan ideals, they put forward educational projects which included practical, 'experimental' subjects as well as the liberal arts. They were particularly concerned with improving the teaching of Latin and Greek. John Dury in *The Reformed School* (London, 1650) complained that at present children are taught *words* before they know *things*. They did not suggest that the study of the classical languages should be abolished, but they hoped 'to introduce a Better, Easier, and Readier Way of Teaching' to overthrow the 'Grammatical Tyranny of teaching Tongues' (Hartlib 1654: 194). Both Hartlib's and Dury's ideas had already been discussed by Petty in his *Advice* of 1648. They demanded a 'plain', simple prose style which would facilitate understanding and could describe the 'reality' or true meaning of things.[6] The point to emphasise is that the Baconian reformers' practical attitude to learning coincided with their Puritan ideals – Puritanism did play a part in their combined educational, religious and social reforms, but not the significant one. Hartlib, Dury, Comenius acknowledged that they owed their inspiration to Bacon's *Instauratio Magna*, 'that most instructive work of the century now beginning'.[7]

When the Royal Society was founded, Hartlib's name was (for political reasons) not mentioned in the official records; he died in 1662, convinced that his work for reform had been wasted. Those, however, in the seventeenth century (and after) familiar with the Baconian spirit

---

[5] In 1642 Hartlib translated Comenius' *Conatuum Pansophicorum Dilucidatio* as *A Reformation of Schools*. Comenius' descriptions of pansophic books, schools, colleges and language are found in *Via Lucis*, written in London in 1641–2 but published in Amsterdam in 1668. For Comenius' relation with the Royal Society see Syfret 1947–8: 115ff.

[6] Arguing for a balanced curriculum George Snell, another associate of Hartlib and Dury, suggested in *The Right Teaching of Useful Knowledge* (London, 1649) that logic, rhetoric and law be taught in the vernacular.

[7] And see Syfret 1947–8: 105, 115.

of reform were well aware of the part that Hartlib played in the foundation of the Royal Society (Robert Boyle acknowledged him as its 'midwife and nurse'). No less important were Hartlib's activities in England's 'Educational Renaissance', for it was Hartlib's realistically orientated education projects based on Baconian principles that served the Dissenters (including Charles Morton, Defoe's teacher) as precedents for their reformed Academies. When in 1662 the first Dissenters' Academies were founded, they drew on the schemes put forward by the Baconian Puritan reformers.[8]

In the present discussion of Defoe's intellectual indebtedness, Hartlib's significance lies in the link he provides between Bacon and the Dissenters' reformed schools of the mid-century. The works listed in the Sales Catalogue of the Defoe/Farewell libraries which have a direct bearing on Defoe's views on trade and education are: Hartlib's *Discourse of Husbandrie used in Brabant and Flanders* (London, 1650) and its enlargement, Hartlib's famous *Legacy* of 1651. The *Legacy*'s basic principle that 'real' knowledge of trades (that is, arts and crafts) ensures economic progress, while ignorance leads to decay and poverty, is an idea that we will find frequently repeated in Defoe's economic works, particularly in his *General History of Trade* (1713) and in *A Tour thro' the Whole Island of Great Britain* (1724–6). The catalogue also lists *The Reformed Spiritual Husbandman* (London, 1652), a collaborative effort of Cressy Dymock, John Dury and Hartlib which concluded with the suggestion for the foundation of Chelsey College.[9]

William Petty (1623–1687)

William Petty and Robert Boyle represent in some ways a bridge between those who saw science as an end in itself, and those who translated the principles of New Science into 'works' and reform. Both Petty and Boyle started their careers as believers in Baconian natural history in the Hartlib group and later made the geographical and intellectual move to the Philosophical Society of Oxford.

William Petty was first educated by the Jesuits in France, subse-

---

[8] For a more detailed discussion of the Baconian Puritans' influence on Morton's Academy see pp. 36–8; Defoe's educational views and projects will be discussed on pp. 55–65 below.

[9] For Defoe's proposal for a foundation of a college in London see pp. 63–4 below.

quently studying medicine at the universities of Utrecht, Leyden and Amsterdam.[10] Upon his return to England in 1646, he became acquainted with Hartlib and his group. Inspired by their ideas for reform and applied science, Petty wrote in 1648 his *Advice*. In this work he proposed the erection of '*a Gymnasium Mechanicum*, or a Colledge for Trades-men' where the study as well as the writing of a Baconian History of Trades could be carried out. Through Hartlib, Petty came into contact with Robert Boyle, and in 1647 the two joined a group of London virtuosi. They met weekly to discuss 'the New Philosophy or Experimental Philosophy' and the group is now recognised as a forerunner of the Royal Society.[11]

By 1648, with the parliamentary re-organisation in Oxford, several members of the London group, including Wilkins, Wallis and Goddard, had moved to Oxford, and Petty followed their example. With this move, Petty left his association with the Baconian Puritan reformers and became more directly engaged in experimental science. In Oxford the group reformed as the 'Philosophical Society'.[12] In 1649, aged 26, Petty was appointed deputy to Clayton, Professor of Astronomy, and one year later received a full professorship. At the same time he was elected to a fellowship at Brasenose College. In June 1650 Petty became a member of the College of Physicians, and also in the same year he was appointed Professor of Music at Gresham College. Despite his success, Petty did not stay very long at Oxford and in May 1652 he accepted the position of Surgeon-General to Cromwell's army in Ireland.

When Petty arrived in Ireland the Civil War had just ended, so he arrived at the time when the Republic's main interest was directed towards an Irish settlement. This in effect meant the repayment of an enormous debt in Irish land, and the first step in this procedure consisted of a survey and a map of all the land available. Petty

[10] The following sketch of Petty's life is based on Aubrey 1949, Evelyn 1959, Fitzmaurice 1895, Masson and Youngson 1960.

[11] See note 1 above and also Dr John Wallis's account of the foundation of the Royal Society given in a letter to Dr Smith of Magdalen College, Oxford, dated 29 January 1696–7. The letter is printed as No. XI of the Appendices to the Publisher's Preface to *Peter Langtoft's Chronicle* in the *Works* of Thomas Hearne (London, 1810).

[12] They first met at Petty's lodgings (as he lived conveniently for their experiments in an apothecary's house) and later, after Petty's departure for Ireland, in Wilkins's rooms at Wadham College.

undertook this formidable task, within the space of thirteen months measuring and recording (on a scale of near 8 miles to the inch) all forfeited land. The 'Down Survey', as it was known, recorded and 'lay'd down' all natural divisions of the country and was completed in the autumn of 1656.

Although the idea of adapting the basic principles of the natural history programme to investigations of social and economic aspects had been contained in Bacon's original plan, it was Petty who first applied it in what he called 'Political Arithmetic'. By analogy with anatomy and arithmetic, it was believed that society could be scientifically studied, 'dissected', and that every part could be known. Eventually, when the entire 'Body Politick' had been explained, it was hoped that social and economic predictive theories could be established. In the Preface to his *Political Anatomy of Ireland* (London, 1691) Petty explained:

> As Students in Medicine, practice their inquiries upon cheap and common *Animals* ... I have chosen *Ireland* as such a *Political Animal* ... 'Tis true, that curious *Dissections* cannot be made without variety of proper Instruments; whereas I have had only a commin *Knife* and *Clout*, instead of the many helps which such a Work requires.
>
> (Petty 1899, I: 129–30)

With his retirement from his public duties in 1659, Petty's interest in experimental philosophy revived, and he renewed his friendship with the members of the 'Philosophical Society' (in the meantime they had moved from Oxford to Gresham College, London). When on 15 July 1662, the 'Royal Society of London for the Improvement of Natural Knowledge' emerged from these meetings and received its royal charter, Petty was named a charter member of its council. He was elected Vice-President in 1674, and in 1684 he founded the Royal Society of Ireland, becoming its first President.

Defoe's references to Petty and to Petty's 'Political Arithmetic', as for example in *An Essay upon Projects* (1697), the *Review*, *Atlas Maritimus and Commercialis; or, a general View of the World, so far as it relates to Trade and Navigation* ... (1728), *A Plan of the English Commerce* (1728), and in his *Tour* make it clear that he was familiar with this new method of exposition. The catalogue of the Defoe/Farewell libraries lists Petty's *Discourse made before the Royal Society 26 November 1674 concerning the* ...

Duplicate Proportions (London, 1674); *Further Observations upon the Dublin Bills, or Accompts of the Houses ... in that City* (London, 1686); *Two Essays in Political Arithmetick concerning ... London and Paris* (London, 1687); *Five Essays in Political Arithmetick – Observations upon the Cities of London and Rome* (London, 1687); and Petty's most important work, *Political Arithmetick* (London, 1691).[13]

Robert Boyle (1627–1691)

The youngest son of Richard Boyle, first earl of Cork, Robert Boyle was educated at Eton and then in Geneva.[14] On his return from the Continent in the 1640s he became acquainted with Hartlib. As indicated above, it was through Hartlib that Boyle published his first work. After a brief stay in Ireland, Boyle in 1656 returned to England to take up permanent residency in Oxford. Joining Wilkins's circle, he re-joined many of his former acquaintances from London.[15] In order to be near the Royal Society, Boyle in 1668 moved to London where his laboratory became a private research centre. During the next twenty-three years, until his death in 1691, he seldom missed a meeting of the Society. When he was elected third President, in 1680, he declined and the place went to Christopher Wren.

Boyle's first major publication was on pneumatics. In 1658 his assistant Robert Hooke helped to construct the 'New Pneumatical Engine', an air-pump (based on a German invention of Otto Guericke (1602–86), Major of Magdeburg), with which Boyle carried out his investigations of the 'Spring' and 'weight' of the air, and the effects of a vacuum (B. Vickers 1987: 48). These experiments led in 1660 to *New Experiments Physico-Mechanical, Touching the Spring of the Air, and its Effects* and in 1662 to the formulation of Boyle's Law (that at constant temperature gas volume and pressure are inversely proportional). In *New Experiments* Boyle faithfully followed Bacon's advice for a plain, un-rhetorical prose style for recording scientific experiments. Boyle was

---

[13] See Appendix p. 180.
[14] The account of Boyle is based on Hall 1965, Fulton 1932 and Fulton 1960. Boyle was an exceptionally prolific writer and only a small selection of his works can be mentioned here. For a full list see Fulton 1961; for the standard edition of the *Works*, see Boyle 1744.
[15] See p. 34 below.

the first major English scientist to write in English and in a simple language that could be read and understood by the uneducated.[16]

Another early work demonstrating Boyle's devotion to Baconian experimental science is *Some Considerations touching the Usefulness of Experimental Natural Philosophy* (Oxford, 1663; vol. II, Oxford, 1671). *Usefulness* contains Boyle's outline for a History of Trades as advocated by Bacon in the *Parasceve*. Regarded as 'the finest apology we have for the History of Trades as an idea', this work will be one of the central texts in the discussion below of the Baconian influence on Defoe's *History of Trade* (Houghton 1941: 57). While being a faithful exponent of Bacon's experimental philosophy, Boyle also developed 'his own private version of Cartesianism' (Fulton 1932: 93). His conception of a corpuscular theory of the nature of matter, first outlined in *The Sceptical Chymist* (London, 1661), was stated in greater detail in *Certain Physiological Essays* (London, 1661) and *The Origine of Formes and Qualities* (Oxford, 1666).

Of the contemporary records documenting Boyle's influence on the development of modern science, Charles Morton's will hold our special attention. In his science lectures, the *Compendium Physicae* (which Defoe would have heard and been asked to copy out), Morton displayed his detailed knowledge of a wide range of Boyle's work. Whether writing on the circumference of the earth, on cold, on fluids, on form, Morton always cites Boyle as the decisive authority. The work that Defoe's teacher seems to have known particularly well was *The Spring and Weight of the Air*.[17]

In view of Morton's knowledge of Boyle's writings, it is not surprising that Defoe should have expressed a similar admiration. Defoe's familiarity with Boyle's works is quite unmistakable. Even if he had not referred to Boyle by name in *The Storm* (1704), *The Consolidator: or, Memoirs of Sundry Transactions from the World in the Moon* (1705), *History of Arts and Sciences* (where Boyle's name is mentioned no less than eleven times) and *The Compleat English Gentleman*, we should still have indisputable proof. In his history of science Defoe discusses the discovery of

---

[16] On Boyle's indebtedness to Bacon see Hall 1950 and Baden Teague, 'The Origins of Robert Boyle's Philosophy' (unpublished Ph.D. thesis, Cambridge University, 1972).

[17] See pp. 40–1 below.

the magnet and loadstone using Boyle's research, and quotes lengthy passages verbatim from *Certain Physiological Essays*.[18]

Boyle combined theology and science, indeed many of his religious writings were off-shoots of his scientific investigations. A defender of natural theology, Boyle studied the wonders of creation convinced that experimental philosophy furthered, and not opposed, Christian faith. Stating his faith in divine providence he declared in *A Free Inquiry into the Vulgarly Received Notion of Nature* (London, 1686):

> It seems to detract from the honour of the great author and governor of the world, that men should ascribe most of the admirable things, that are to be met with in it, not to him, but to a certain nature.
>
> (Boyle 1744, IV: 361)

That Defoe knew of the 'Providence' tradition is clear from a number of his works beginning with *The Storm* and in 1706 *Caledonia, a Poem in Honour of Scotland, and the Scots Nation*. The most explicit working out of the theme is in part three of *Robinson Crusoe* (1720), where Defoe offers his own definition of Providence. The fact then that *A Free Inquiry* (which was with Sprat's *History of the Royal Society*, the seventeenth-century propaganda for Baconian science) was one of the four works by Boyle in the Defoe/Farewell libraries should not be overlooked.[19]

Robert Hooke (1635–1703)

Robert Hooke was educated at Westminster School and Christ Church, Oxford. Indications of his 'Mechanical Genius' were soon evident and he was engaged as scientific assistant first by Thomas Willis and then by Robert Boyle.[20] In 1658 while assisting Boyle in experiments concerning 'the Spring and weight of air', he devised the air-pump. Hooke was appointed Curator of Experiments of the Royal Society in 1662 and held this position for the next forty years. His duty was to furnish the Society for every sitting, that is once a week, 'with three or four considerable Experiments', and he fulfilled this de-

---

[18] See pp. 76–7 below.
[19] For a more detailed discussion of this complicated issue see p. 69 and pp. 112–20 below. For a list of the works by Boyle in the catalogue of the Defoe/Farewell libraries see the Appendix pp. 177 and 180.
[20] The material relating to Robert Hooke's life is mainly taken from Hooke 1705, Gunther 1920–67, 'Espinasse 1956 and Andrade 1960.

manding task with unfailing resourcefulness and industry. When Sir John Cutler in 1664 founded a lectureship in mechanics, Hooke was its first holder; in 1665 he became Gresham Professor of Geometry. In 1677 Hooke succeeded Oldenburgh and became Secretary to the Society, so filling for some time the offices of both Secretary and Curator.

During the plague of 1665 Hooke accompanied Wilkins and Petty to Surrey. John Evelyn visited them, recording in his diary that he had found them

> contriving Charriots, new rigges for *ships*, a Wheel for one to run races in, & other mechanical inventions, & perhaps three such persons together were not to be found else where in Europ, for parts and ingenuity.
>
> (Evelyn 1959: 479)

With the Great Fire of London a new opportunity for Hooke's many-sided genius offered itself. He became City Surveyor, and as Christopher Wren's chief assistant designed among other buildings Bedlam Hospital, the Royal College of Physicians, and the Monument.

Hooke was blessed with an extraordinarily fecund mind. Besides the air-pump, his inventions included an anchor-escapement-mechanism for the pendulum clock, the wheel-barometer, a clock-driven telescope, a sealed thermometer, a weather-clock (the first to offer prognoses determined by measurable physical causes), a universal joint, and a hygrometer. In a lecture entitled *An Attempt to prove the Annual Motion of the Earth* (London, 1674), Hooke demonstrated that 'the Power of any Spring is in the same proportion with the Tension thereof', now known as Hooke's Law. He lectured on optics, light, colour, heat, meteoreology, astronomy, gravity, combustion and comets. He was one of the first to reject the previously accepted argument that fossils were mere '*lusus naturae*' (sports of nature) and the product of some 'Plastick faculty inherent in the Earth'. Basing his argument on Baconian scientific observation and classification, Hooke argued that fossils were petrified natural bodies. Hooke's most significant contribution to the development of modern science was in the field of microscopy. In 1663 the Society asked him to produce at every meeting at least one microscopical observation. Hooke delighted the Fellows with enlargements of mosses, the pores of cork, the wings of flies, the head of an ant, the edge of a razor, the teeth of a snail, etc. When in 1665 these experiments were collected in *Micrographia*, Hooke supplied the most accurate illustrations of these microscopical objects the world had ever seen.

In 1705 Richard Waller published the *Posthumous Works of Robert Hooke*, which is listed in the catalogue of the Defoe/Farewell libraries. It contains lectures and discourses on the petrification of stone and wood, the origin of fossils, the cause and effect of comets and earthquakes. In other words, it deals with subjects that particularly interested Defoe and on which he expressed an opinion right up to the end of his life. Hooke's *Posthumous Works* also contains a detailed outline for a Baconian natural history of trade (Hooke 1705: 57ff and see pp. 84 and 103 below).

John Wilkins (1614–1692)

John Wilkins was of Puritan upbringing. He trained for the ministry and after graduation from Magdalen Hall, Oxford, became chaplain to the exiled Prince (Elector) Palatine. His main ecclesiastical appointments thereafter were Dean of Ripon (1660–72), vicar of St. Lawrence Jewry in the City of London (1663–8) and Bishop of Chester from 1668 until his death in 1692.[21] Besides his career in the church, Wilkins had an equally successful academic career. He was a parliamentarian, and when Oxford came under parliamentarian control in 1648, he was made Warden of Wadham. In 1659 he moved to Cambridge as Master of Trinity; ejected at the Restoration, he went to live in London. When in 1660, after a lecture by Christopher Wren, the foundation of 'a college for the promoting of physico-mathematical experimental learning' was discussed, Wilkins chaired the meeting. Until the Society received its royal charter, a president was elected monthly, Wilkins holding the position on four occasions. He regularly attended its meetings, and was on all the important committees; he became a member of the council and one of its two secretaries.

In 1664 it was decided that the Society's undertakings should be carried out by eight sub-committees, and Wilkins became a member of

---

[21] For the following biographical sketch of Wilkins, I have consulted Stimson 1931, Bowen and Hartley 1960, Crowther 1960 and Shapiro 1965.

five of them: committees for the History of Trade, Mechanics, the 'Georgical' (that is, Agricultural), Anatomy and Correspondence (Stimson 1931: 552). Minutes of these committees are still extant and prove that the Fellows discussed and outlined histories of trades, agriculture and the weather. Wilkins's recommendation for a history of the weather may be directly linked to at least two projects of this time: Hooke's 'Method for Making a History of the Weather' and Christopher Wren's 'History of Seasons' (Sprat 1959: 173–9, 312–13).[22] This subject will be relevant to Defoe's 'Observation of the soil and climate of the continent of America' in *A New Voyage*; in *Crusoe* 1 Defoe lets his hero meticulously observe and table the changes of the weather.

Wilkins was less a practising scientist than a populariser of the experimental method. In *The Discovery of a World in the Moon* (London, 1638) and *A Discourse Concerning a New Planet* (London, 1640) he put forward the sensational idea that it might be possible to fly to the moon. In *Mercury, or the Secret and Swift Messenger* (London, 1641), an outline of a cryptographic system, Wilkins pre-echoed his later concern with a 'real character' or artificial language. Also belonging to his university days is *Mathematical Magick; or, the Wonders that may be performed by Mechanical Geometry* (London, 1648). Wilkins here explained the principles and practical uses of the balance, lever, wheel, pulley, wedge and screw. The 'automata' discussed include 'a chariot with sails to catch the wind', and 'flying chariots'.

Among Wilkins's many interests was his creation of a new, philosophical (that is, scientific) language. His *Essay Towards a Real Character and a Philosophical Language* (London, 1668) set out to replace our conventional language with characters that could express the reality of things (hence the 'real character'). Taking up Bacon's advice, Wilkins's universal language scheme carried to the logical conclusion the experimentalists' stress on things. The scheme was impractical and was regarded by many as an embarrassing failure. It is, however, important to note that the inspiration for the 'real character' emanated from the Baconian demand for a precise, plain mode of expression.

Wilkins's main contribution to science consists in his attracting to

---

[22] John Locke's registers of the weather, kept between June 1666 and October 1682, were incorporated into Boyle's *General History of the Air* (Boyle 1744, V: 136–61, and see pp. 70, 106–9 and 136 below).

Wadham the outstanding scientists of his day: Ward, Rooke, Wallis, Petty, Bathurst, Willis, Boyle, Wren, Hooke, Sprat and others. At Oxford, and specifically at Wadham, Wilkins created an intellectual milieu in which experimental learning could flourish; the foundation first of the Oxford Philosophical Society and then of the Royal Society of London are to a large extent due to his knowledge and efforts.

The name of one other 'intruded' alumnus must be mentioned here: Charles Morton, Defoe's teacher, was at Wadham from 1649 to *c.* 1655. As we will see soon, during his time at Oxford Morton was exposed to the New Sciences; when Morton founded his Academy for Dissenters, he had a stock of experimental learning and practice to draw on. Defoe shows his familiarity with Wilkins's projects in two early works: *An Essay upon Projects* of 1697 and *The Consolidator* of 1705.[23] The catalogue of the Defoe/Farewell libraries includes *A Discourse concerning the Gift of Prayer*, Wilkins's tractate on the art of prayer. The relevance of *Gift of Prayer* will become clear in the following chapter, discussing Charles Morton's use of Wilkins's advice for plain preaching and prayer.[24]

---

[23] For Defoe's display of his knowledge of Wilkins see pp. 70–2 below.
[24] On Morton's indebtedness to Wilkins see pp. 48–50 below.

3

Charles Morton and the New Sciences

Charles Morton (1627–1698)

Charles Morton was a scholar at Wadham College, Oxford, from 1649 to *c.* 1655. It is recorded that he excelled himself in mathematics, 'especially the Mechanic part thereof', and that he was 'extremely valued by Dr. Wilkins for his mathematical genius'.[1] At first a Royalist, Morton became a Puritan 'when he found that the laxest members of the university were attracted to that side', that is, the Royalists (*DNB*). He trained for the ministry and left Oxford *c.* 1655 for a rectory in Cornwall, but was ousted when the Act of Uniformity was renewed in 1662. Little is known of his movements during the next few years; it is believed that he opened his Academy for Dissenters in Newington Green soon after 1662 (Matthews 1934: 356–7). Defoe was a student at this Academy from *c.* 1674 to *c.* 1679.[2]

By founding a school of higher education Morton broke the so-called 'Stamford-Oath' (which forbade graduates of Oxford and Cambridge to teach outside the two main universities) and was charged with perjury. For many years he held out, but when in 1685 he was invited to a post in New England he decided to accept. At Harvard, Morton taught the methods and goals of experimental science, his reputation as

---

[1] See Wesley 1704: 5, Calamy 1775, I: 274 and Matthews 1934: 356–7.
[2] See Lee 1869, I: 14, McLachlan 1931: 26, Girdler 1953: 574 and Bastian 1981: 48.

an outstanding teacher being so great that in 1697 he was appointed Vice-President of Harvard College. In the words of his modern biographer, 'with Morton's help Harvard College pulled out of the bog of medieval Science, and set her face toward experimental philosophy, and the "century of enlightenment"' (Morison 1940: xxiii).

Morton's scientific education at Oxford

Although the curriculum at Oxford during the middle of the seventeenth century was outmoded, a student could, if he desired, become familiar with the principles of the New Sciences.[3] Scientific activity took place on three levels. First, in the statutory lectures and disputations; second and more importantly, in an unofficial form in the gatherings of scientifically orientated members of the university; finally, through the expanding book trade. Answering John Webster's fierce attack on the universities, John Wilkins and Seth Ward pointed out that 'those that understand these places, do know that there is not to be wished a more generall liberty in point of judgment or debate'. 'There is scarce any Hypothesis', wrote the defenders of the present system, but it 'hath here its strenuous Assertours ... [in] the Atomicall and Magneticall ... Philosophy ... Witnesse the publick *Lectures of our Professors* ...'[4]

By the middle of the century Oxford could boast several substantial scientific benefactions: the Savilian Professorships of Geometry and Astronomy (endowed in 1619), the Sedleian Professorship of Natural Philosophy (1621), the Earl of Danby's Botanical Garden, where from the middle of the 1620s observations and experimental studies could be carried out, and the Tomlins Readership in Anatomy, founded in 1624. During Morton's stay at Oxford Sir William Petty was Tomlins Reader and John Wallis held the Savilian Professorship of Geometry. It is very unlikely that such keen advocates of Baconian science as Petty and Wallis would not have communicated their excitement and introduced the new ideas into their teaching. Although the statutory

---

[3] For the following account of education at Oxford in the middle of the seventeenth century, I am indebted to Frank 1973 and 1980, and Feingold 1984.
[4] See John Webster 1654: 20, 41–2, 98, 105–6; Ward and Wilkins 1654: 2. For a more recent treatment of the debate see Debus 1970.

university curriculum was rigid as to the form and length of the lectures, no strict recommendations seem to have governed their actual contents (Frank 1973: 200). In spite of the continuity of Laud's Oxford Statutes of 1636 the adherents to the experimental science could, if they wanted to, let their ideas infiltrate into their teaching.

In addition to the formal lectures, there emerged in the 1640s and 1650s a number of unofficial scientific gatherings. Of these, Wilkins's group at Wadham was the most distinguished, and in the history of English science the most significant. It was during Morton's time that this group, the Oxford Philosophical Society, changed its venue from Petty's lodgings to Wilkins's in Wadham. A list of its most prominent members will indicate the scientific ambience to which Defoe's teacher was exposed. Besides Wilkins, there was Seth Ward (mathematician and astronomer), Sir Christopher Wren (scientist and architect), Lawrence Rooke (astronomer), John Wallis (mathematician), William Petty (anatomist and economist), Jonathan Goddard (physician and maker of telescopes), Walter Pope (astronomer) and Robert Boyle (physicist and chemist). With the exception of Ward, Wallis and Pope, these scientists were among the twelve who on 28 November 1660 met at Gresham College to discuss the foundation of the Royal Society of London.

Robert Boyle has been described as Morton's 'chief scientific master' (Hornberger 1940: xxxix), but this claim needs qualifying. Boyle, an aristocrat, did not hold a teaching post at Oxford. He visited Wadham on a number of occasions in the early 1650s and came to live in Oxford as late as 1654 or 1655, that is, when Morton was about to leave. The indebtedness to Boyle – and there is no doubt that Boyle was Morton's main source of inspiration – is based neither on formal nor informal lessons, but on Morton's detailed knowledge of Boyle's written work. Boyle was exceptionally productive, and with the publication of forty-nine works he contributed more than anyone else to the scientific book trade (Frank 1980: 44). Both the university and the college libraries would have given easy access to these writings. Morton, it seems, was sufficiently committed not only to make use of the intellectual facilities of the university but to pursue experimental activity on his own, with the help of books.

Oxford, much more than Cambridge, had by the middle of the

century a considerable collection of mathematical and mechanical instruments, and natural history specimens. By far the greatest collection of these was to be found in Wilkins's lodgings. John Evelyn recorded on 13 July 1654 that he had

> din'd, at that most obliging & universaly Curious Dr. *Wilkins*'s, at Waddum, who was the first who shew'd me the *Transparant Apiaries*, which he had built like *Castles & Palaces* & so ordered them one upon another, as to take the *Hony* without destroying the *Bees*; These were adorn'd with variety of *Dials, little Statues, Vanes &c:*... He had also contrivd an hollow Statue which gave a Voice, & utterd words, by a long & conceald pipe which went to its mouth, whilst one spake thro it, at a good distance ... He had above in his Gallery & Lodgings variety of *Shadows*, Dyals, Perspectives, places to introduce the *Species*, & many other artificial, mathematical, Magical curiosities: A Way-Wiser, a *Thermometer*; a monstrous *Magnes, Conic & other Sections*, a Balance on a demie Circle, most of them of his owne & that prodigious young Scholar, Mr. *Chr: Wren*, who presented me with a piece of *White Marble* he had stained with a lively red very deepe, as beautifull as if it had been naturall.[5]

There is no doubt that a student living at Wadham and having scientific interests, would have been influenced by the intellectual excitement around him. At Wadham, Morton was first made aware of the New Sciences; it was here that he heard of the experimental scientists, their activities and inventions. When Morton opened his own academy and came to write his science lectures, he could apply what he had observed and learnt in his student days at Oxford.

Morton was not destined to become a 'major' scientist. As far as we know, he was not an official member of the 'philosophical clubbe', nor did he later become a Fellow of the Royal Society.[6] His talents were not such that he could contribute to the solving of scientific problems, yet he is to be counted among those followers of experimental science who through dissemination of the new ideas played a significant role in the advancement of learning.

---

[5] Evelyn 1959: 341. Wilkins's collection of mathematical and mechanical instruments was a direct response to Bacon's advice given in *New Atlantis*, Bacon 1857, III: 162–3. For Defoe's knowledge of some of these inventions see pp. 70–2 below.

[6] Morton contributed once, with the assistance of his close friend and Boyle's sometime assistant, Daniel Cox, to the *Philosophical Transactions*; the paper was entitled 'The Improvement of Cornwall by Sea-Sand' and appeared 26 April 1676.

Dissenting Academies

The founders' main consideration was to reproduce in their Academies the standard of education of the traditional universities. They not only succeeded, but in some fields outstripped the orthodox institutions. Due to their high standards and due to the fact that they opened their doors to anyone who sought higher education, the Dissenting Academies quickly began to flourish. Morton's, which is usually classified with the first group of Academies (in existence from 1662 to 1685), was one of the best in the country.[7]

Although the tutors varied in how they replaced, modified and extended the orthodox curriculum, they all stressed a practical approach to learning. One example of their utilitarian attitude was their decision to include natural science in their teaching. Morton was among the very few of the earliest tutors who introduced the experimental approach into the lecture-hall, and who made reference to recent research. Samuel Wesley (who was one generation after Defoe at the Newington Green Academy) described Morton's Academy as

> the most considerable [in England], having annext a fine Garden, Bowling-Green, Fish-Pond, and within a Laboratory, and some not inconsiderable Rarities, with Air-Pumps, *Thermometers*, and all sorts of Mathematical Instruments.
>
> (Wesley 1704: 6)

This would indicate that Morton reconstituted Wilkins's Oxford garden in his London suburb, making personal observation and experimentation part of his science course. We know that he taught that knowledge without practice was no knowledge; 'solid inferences', Morton declared, must be based on 'well observed Experiments'.

Among his contemporaries, Morton was distinguished for the place he allotted to modern languages. In his eagerness to rejuvenate the outmoded curriculum Morton took still more drastic measures, in that he delivered all his lectures in English and made English itself a subject to be studied. In what is almost certainly a description of Newington Green, in 1712 Defoe described a school in which

---

[7] For discussions of the Dissenters' Academies see Parker 1914; McLachlan 1931; J. W. A. Smith 1954; M. Hunter 1975 and 1981.

the Master or Tutor ... read all his Lectures, gave all his Systems, whether of *Phylosophy or Divinity*, in *English*; had all his Declaimings and Dissertations in the *English* Tongue. And tho' the Scholars from that Place were not Distitute in the Languages, yet it is observ'd of them, they were by this made Masters of the *English* Tongue, and more of them excelled in that Particular, than of any School at that Time.

(*The Present State of the Parties in Great Britain*, p. 319)

Morton was not consistently 'modern', but rather a mixture of old and new ideas. His science lectures, which will be discussed below, are proof of this, as is a little tractate, entitled *The Spirit of Man*, which outlines a balanced curriculum containing theoretical and practical science, and also the liberal arts (Morton 1693: 21–2). Reading discussions of the earliest Dissenters' Academies is to be frequently reminded of their remarkable originality and modernity in anticipating our education today. However, hardly any aspect of the Dissenters' programme of educational reform was in fact new. Morton's stress on the usefulness and applicability of knowledge, the introduction of experimental science, his wish not to abolish Latin and Greek but to improve its teaching, finally, his decision to teach in English – all of this had been adumbrated by the Baconian Puritan reformers. Hartlib, Dury, Petty were committed to introducing 'a Better, Easier and Readier Way' to teach the classical languages, and 'to make use of that which they knew for the benefit of Mankind' (see p. 21 above). If Charles Morton held that a truly liberal education should include utilitarian and empirical subjects, so did the first-generation Baconians. As early as 1648 Wilkins had written in *Mathematical Magic* that 'those other disciplines of Logick, Rhetorick, etc. doe not more protect and adorn the mind, than these Mechanicall powers doe the body' (p. 10).

Although the year of the Conformity Legislation which debarred the Dissenters from the universities is the year from which the Academies date, the principles upon which these were founded antedate 1662 by twenty or more years. Intellectually, the founders of the nonconformists' schools belong to a movement that first gathered momentum in 1645, when Hartlib and his circle met to discuss educational reform. As I have tried to show, Hartlib and his associates based their programme of reform upon a fusion of Baconian and Puritan ideals. Their proposals for founding practically orientated colleges came to an abrupt end with the Restoration, yet the halt to England's educational revolution was only temporary, for the Dissenters soon took up and 'kept alive in their Academies the spirit of Hartlib and those who had worked with him in the spread of realism' (Parker 1914: 44, Jones 1961: 174 and Hill 1965: 109).

I am not arguing for a direct (personal) link between Morton and the Hartlib circle, but I would like to suggest that Morton was motivated by the same general scientific-religious ideals which had inspired the Baconian Puritan reforms. Bacon's programme for the reform of knowledge is the decisive and unifying element connecting Morton with the Puritan reformers of education. Such statements as that Morton had no 'experience of any university in which similar work was being done' and that his 'activities must therefore be regarded as an original contribution', must be read with utmost care (J. W. A. Smith 1954: 247). Morton was the first to introduce practical science work into the classroom of the Dissenting Academies, but he had had experience of applied experimental science during his time at Oxford. Through his acceptance of the tenets of the New Sciences and through his dissemination of these ideas, Morton stands in direct line of descent from Bacon.

Morton's Compendium Physicae

We are extremely fortunate in possessing a school-text written by Morton for his students at Newington Green. When Morton arrived in New England in 1686 he brought with him a sequence of science lectures which 'was adopted at Harvard as a textbook in Physics from 1687 onwards'. As the *Compendium Physicae*, Morton's lectures 'remained the foundation of instruction in Natural Science at the College for some forty years' (Morison 1936, pt. I: 238). We know that Defoe considered Morton's 'Manuscripts of Science' – which almost certainly is the *Compendium* – so precious that he preserved them for more than twenty-five years, remarking in 1704, when he defended Morton's Academy against Wesley's violent attack, that

> the Author of these Sheets happens to be one that had, what little Education he can pretend to, under the same Master that Gentleman

was Taught by, *viz.* Mr. *Charles Morton of Newington Green* and I have now by me the Manuscripts of Science, the Exercises and Actions of his School ...

(*More Short-Ways with the Dissenters*, pp. 5–6)

The compendious systems of nature and the 'severall Arts and Sciences' give us insight into Morton's concept of reality, and into the way in which he communicated his perceptions to his students. 'Anyone interested in Defoe's views on science should read the *Compendium Physicae*', a scholar wrote in 1953 (Girdler, 1953: 587): not many seem to have done so.

Even a glance through the *Compendium* will show that Morton was abreast of much recent scientific research. Refuting the ancient theory that lower animals breed from putrefaction, Morton quoted from William Harvey's work in embryology, *Exercitationes de Generatione Animalium* (London, 1651); discussing the claims for astrological influence on human life and dismissing them, he quoted from Dr John Spencer's 'Discourse concerning Prodigies' published in 1663. In his remark that 'the texture of Every feather gives in the Microscope a sight most astonishing' Morton made reference to Hooke's *Micrographia* of 1665 (*Compendium*, Morton 1940: 139, 146–8, 93, 191). The chapter 'Of Water', the most modern part of the work, shows Morton's acquaintance with hydrodynamics: he describes the 'Water-Screw' or Archimedean screw, the siphon, the double pump, a 'water engine or springer', that is, the hydraulic ram, and others (pp. 58–60). Speculating on why the sea is salt, he appears to have been familiar with a letter by Martin Lister which appeared in the *Philosophical Transactions*, 20 February 1683–4. In his definition of elasticity Morton noted that

> the reason of Elasticity by Dr. Petty is Judged to be a magnetick virtue in Each of the particles whereby they incline to stand in a line so as their amicable poles may touch each other; and therefore if they are removed from that verticity they will of their own accord [recurr therunto]. I know not a better Solution of this P[h]ainominon.
>
> (p. 39)

Morton's references always help to make or illustrate a point, and they are always based on intimate knowledge of the experimentalists' work. His explanation of the 'Whispering Place at Glosester' is interesting on two accounts: first, Morton for his argument combined personal observation and experimentation with a long description of the 'Whispering Place' delivered to the Royal Society on 5 November 1662; second, almost five decades later, Defoe followed in Morton's footsteps and used his teacher's experimental approach to explain the 'mystery' of the echo in Gloucester cathedral.[8]

In his research on the ebb and flow of the sea Morton made extensive use of John Wallis's essay 'exhibiting his Hypothesis about the Flux and Reflux of the Sea', published in the *Philosophical Transactions*, 6 August 1666.[9] The way in which Morton used, reduced and simplified Wallis's essay to make it more easily understandable for his students indicates his grasp of the material and tells us something of his teaching skill. In the chapters 'Of Comets' and 'Of Procreation' much is culled from Robert Hooke's research published in his *Lectiones Cutlerianae* (London, 1679). Again, chapters 'Of Air' and 'Of the World' use extensively Henry Power's *Experimental Philosophy* (London, 1664). Other experimental scientists could be listed, but the author most frequently referred to and quoted from is Robert Boyle. Of the thirty-one chapters in the *Compendium* no less than eleven draw heavily on Boyle's work. The work that Morton made most use of was Boyle's famous *The Spring and Weight of the Air* (Oxford, 1660).[10]

Morton's use of the works of the experimental scientists was based on careful study of their method and aims. Closer investigation shows a remarkable correspondence between the areas of research of Morton and the Baconian scientists. A list of shared interests includes: cold, heat, clouds, rainbows, the cause of fire, rarefaction, magnetism, gems and gold, the elasticity of the air, effluvia, urine, blood, among others. The congruence of Morton's and the New Scientists' subjects of investigation cannot be dismissed as accidental. When Bacon died he had left his history of nature unfinished, but his writings provided his

---

[8] Morton 1940: 167, and see Defoe's use of Morton's experimental observations in *A New Voyage* p. 149, and the *Tour*, p. 159, below.

[9] Nearly one-third of Morton's chapter 8, 'Of Water', is derived from Wallis's essay (Hornberger 1940: 62n). Morton also made a synopsis of the essay and inserted it as an appendix to the *Compendium*.

[10] Morton's chapters 2, 6, 7, 8, 12, 15, 17, 18, 19, 20, 23 are based on Boyle's writings. Ideas discussed in Boyle's *Spring and Weight of the Air* helped Morton to write chapters 8, 19, and 24 entitled 'Of Air', 'Of Animate Bodyes in Generall and Speciall Vegetable and Sensable', 'Of Hearing'.

disciples with clear directions to complete the natural history programme. Bacon had urged his disciples to collect and order their experiments of nature into 'histories' (that is, systematic studies), and he had provided them with long lists of titles. The work in which Bacon outlined in greatest detail his idea of making 'histories' is the *Parasceve, or Preparative towards a Natural and Experimental History*, affixed to the *Novum Organum* (London, 1620). This work concludes with 130 suggested 'histories' or areas of investigation, the 'Catalogue of Particular Histories by Titles'. Responding to Bacon's advice, his successors compiled histories under the suggested headings. Thus, Wallis wrote a piece on 'the Flux and Reflux of the Sea' in accordance with Bacon's title no. 20, proposing a 'History of Ebbs and Flows of the Sea'; Morton's chapter 9, 'Of Water', which discusses the 'phenomena of tides' and uses Wallis's work, is a continuation of this Baconian investigation. Bacon's titles nos. 10, 11 and 12 give directions for histories 'of Showers, Ordinary, Stormy, and Prodigious; of Hail, Snow, Frost, Hoar-frost, Fog, Dew; of all other things that fall or descend from above', and no. 14 advises on a 'History of Air as a whole, or in the Configuration of the World'. Following this appeal Boyle designed and began *The General History of the Air* and entitled his sections 42 to 46 respectively: 'Of Dew', 'Of Rain', 'Of Hail', 'Of Snow' and 'Of other Things falling out of the Air'. Morton's study of air, rain, dew, hail and snow is to be found in his chapters 8, 9 and 15 entitled: 'Of Air', 'Of Water' and 'Of Watery Comets'. Bacon had directed that there be histories of earth and sea, of fiery meteors, of comets; Morton used these suggestions and named his chapters 10, 12 and 13 'Of Earth', 'Of the Species of Mixt Bodyes, and of Fiery Meteors', and 'Of Comets'. A special section was given by Bacon to the 'History of Man', and he recommended that histories of vision, of hearing, of smell, of taste, and of touch be written; all of these find precise parallels in chapters in the *Compendium* entitled 'Of Seeing', 'Of Hearing', 'Of Smelling, Taste and Touch'.[11]

More examples could be given but these must suffice to show the direct influence of Bacon upon Morton's science lectures – a point that seems to have gone unnoticed so far. Critics have recognised the

---

[11] I have learnt from Baden Teague's 'The Origins of Robert Boyle's Philosophy' (unpublished Ph.D. thesis, Cambridge University, 1972).

Boylean influence upon Morton's *Compendium*, yet it must be stressed that, before Boyle, the Baconian experimental philosophy gives this work its structure and its methodology (this despite Morton's occasional reference to and reliance on the 'ancients'). In true Baconian spirit Morton believed that comprehensive systems 'of all severall Arts and Sciences' would be

> useful, partly, for the instructing of youth in schools and academies; and partly, that men may have, from time to time, an inventory of what hath been already discovered; whereby the needless labour of seeking after known things may be prevented, and the progress of mankind, as to knowledge, might the better appear.

These are Boyle's words prefacing his *Certain Physiological Essays*, but the spirit that gives them life and meaning is Bacon's – indeed Boyle would have been the first to acknowledge his indebtedness to Bacon, 'the great architect of experimental philosophy' (Boyle 1744, I: 193; Bacon 1857, IV: 255). Morton adopted Bacon's philosophy for the direct inspection of nature, for method and for the establishing of theories. The *Compendium* in its collection and division of knowledge, in its stress on personal observation and personal experimentation uses and *teaches* the fundamental principles of the New Sciences. Morton's 'Manuscripts of Science' taught a system of thought that contributed to forming Defoe's mind.

Science and language

One cannot discuss the Baconian reform of science and education without referring to the related important issue of the time, the relation between science and language.

The moderate and immoderate attack on words

The distrust of language and the concomitant demand for plain prose which can be detected in the middle years of the seventeenth century, had three main sources: first, there was a general stylistic reaction against the extravagance of Elizabethan and Jacobean language; second, the influence of Bacon's experimental philosophy; and finally, the parallel movement in the church which advised that imaginative flights and stylistic embellishments be replaced with a new, pure style. These three movements overlapped to the extent that all three contrasted honest plainness with deceitful rhetoric.[12] I am mainly concerned with the second, the influence derived from the characteristic features of experimental science.

Bacon's reform of science was based on first-hand observation and gathering of data, calling for a style that could describe nature as 'she is in fact'. Outlining the principles for collecting observations of nature he advised:

> And for all that concerns ornaments of speech, similitudes, treasury of eloquence, and such like emptinesses, let it be utterly dismissed. Also let all those *things* which are admitted be themselves set down briefly and concisely, so that they may be nothing less than *words*.
> (Bacon 1857, IV: 254, my italics)

Bacon's philosophy was a philosophy of things not of words. To understand the distinction between words and things we have to go back to classical rhetoric, where the phrase was couched in the formula *res et verba*, demanding an equal stress on both aspects of expression. Thus Cato the Elder recommended: '*Rem tene, verba sequentur*' ('look after the subject-matter, and the words will look after themselves'); similar advice can be found in Cicero, Seneca and Horace.[13] This is the tradition that Bacon followed when he advocated a balanced study of observation and description, and when he attacked those who 'hunted more after words than matter' (III: 283).

It should be noted that Bacon made his demand for plain prose within the context of describing the principles of his natural history programme. Only the scientist was required 'to set down things briefly and concisely': in all other spheres of communication Bacon continued to use the skills of rhetoric and metaphor. Following Bacon's plea for plain prose, his disciples inveighed against unnecessary scholarly citations and rhetorical flourishes. The new style, John Wilkins insisted, must persuade by its simplicity and integrity; it 'must be plain and

---

[12] See Jones 1951, a collection of essays. For a more recent discussion of the seventeenth-century reform of prose see B. Vickers 1985: 1–76.

[13] In his valuable essay A. C. Howell shows that in the course of the century the term *res*, meaning subject-matter, became confused with *res* meaning *things*, and that there can be discerned a tendency which assumed 'that *things* should be expressible in *words*, or conversely, *words* should represent *things*, not metaphysical and abstract concepts', Howell 1946: 131.

naturall, not being darkned with the affectation of Scholasticall harshnesse, or Rhetoricall flourishes'. 'Obscurity in the discourse', Wilkins was convinced, was an 'argument of ignorance in the minde', while 'the greatest learning is to be seen in the greatest plainnesse' (Wilkins 1646: 72). Robert Hooke, warning against the misuse or abuse of language, pointed out that only such words should be used which can 'express the Matter with the least Ambiguity, and the greatest Plainness and Significancy ... avoiding all kinds of Rhetorical Flourishes, or Oratorical Garnishes' (Hooke 1705: 63). And Robert Boyle, another advocate of concise expression, remarked: 'I should think myself guilty of a very Childish vanity, if the use I made of Language were so to write as to be the lesse understood.' Clearly, the purpose of expression is to express and communicate, not to obfuscate. In the preface to *Certain Physiological Essays* Boyle declared that in scientific discourses

> where our design is only to inform readers, not to delight or persuade them, perspicuity ought to be esteemed at least one of the best qualifications of a style; and to affect needless rhetorical ornaments in setting down an experiment, or explicating something abstruse in nature, were little less improper, than it were (for him that designs not to look directly upon the sun it self) to paint the eye-glasses of a telescope, whose clearness is their commendation, and in which even the most delightful colours cannot so much please the eye, as they would hinder the sight.
>
> (Boyle 1744, I: 195)

Again and again the compilers of Baconian natural histories insisted upon the value and necessity of a simple, plain style. In contemporary discussions of the old and new style, a whole series of distinctions was commonly made: while the philological methods of the 'ancients' are nothing but 'vapours' and 'laborious webs', the reformed style conveys the strength and significance of the argument; the old style is 'useless', the new is 'useful'; the first gives merely the 'shells' and 'husks', the latter presents the 'kernel' and 'fruit' of learning; or again, the 'dress', 'colouring' and 'false paint' of rhetoric is juxtaposed with the style that can express the truth and reality of things. Debunking the 'intolerable ambiguity' of the schoolmen, Boyle prescribed a style that was 'like a diamond, as well very clear as perfectly solid' (*Sceptical Chymist*). 'Solid', 'clear', 'real', 'fruitful', 'honest', these are some of the terms synonymous with the plain prose style.

The Baconians argued that it was in the common, natural way of speaking (that is, not in the language of the elite, the 'wits and scholars') that most common sense and experience was to be found. Ideally a thing-like, mathematically plain and therefore unambiguous correspondence between words and things should be aimed at. Taking up Bacon's suggestion that we 'imitate the method of the Mathematicians', Hooke proposed a 'Geometrical Algebra' wherein 'many and very perplex Quantities by a few obvious and plain Symbols' could be expressed (Hooke 1705: 64–5; cf. Bacon 1857, III: 396–7).

When Thomas Sprat in his *History of the Royal Society* (London, 1667) described the Fellows' manner of conducting their spoken and written proceedings, he worked under the careful direction of John Wilkins (Sprat was not a scientist but a clergyman; he later became Bishop of Rochester). As has been recognised, Sprat's official description of the Society's *'manner of Discourse'* directly reflected Wilkins's ideas about style and expression. Wilkins more than anyone else was instrumental in determining the Society's standard of prose (see Stimson 1931, and Christensen 1946). The fact that Sprat's *History* represents Wilkins's attitude to prose is of some importance. The relevant passage from the *History* is frequently quoted in analyses of Defoe's prose style and it has been suggested that there is a connection between Defoe's and Sprat's (or rather Wilkins's) attitude to prose. Since, as will be shown below, Charles Morton based his advice on plain preaching on Wilkins's work, Morton may well be the direct link in this connection.

The passage in Sprat's *History* reads:

> there is one thing more, about which the Society has been most sollicitous; and that is, the manner of their Discourse: which, unless they had been very watchful to keep in due temper, the whole spirit and vigour of their Design, had been soon eaten out, by the luxury and redundance of speech. The ill effects of this superfluity of talking, have already overwhelm'd most other *Arts* and *Professions*.
>
> (p. 111) [14]

Becoming heated in his argument, Sprat then went on to attack the use

---

[14] Compare this with Wilkins's fear that 'the grand imposture of Phrases hath almost eaten out solid Knowledge in all professions', Wilkins 1668: 18.

of rhetoric and metaphor, concluding that '*eloquence* ought to be banish'd out of all *civil* Societies, as a thing fatal to Peace and good Manners'. Describing the remedy that the Society has been resolute in putting into execution, he stated that the Fellows have returned to a

> primitive purity, and shortness, when men deliver'd so many things, almost in an equal number of words. They have exacted from all their members, a close, naked, natural way of speaking; positive expressions; clear senses; a native easiness: bringing all things as near the Mathematical plainness, as they can: and preferring the language of Artizans, Countrymen, and Merchants, before that, of Wits, or Scholars.
>
> (p. 113)

Two points need to be made. Sprat – or, rather, Wilkins – in his demand for a precise, 'close, naked and natural way of speaking', followed closely in Bacon's footsteps. Like Bacon, they distinguished between the language of literature and the language for reporting scientific experiments.[15] It was only in the latter, in scientific discourses, that rhetoric and metaphor were jettisoned as potentially subversive of rational thinking and writing. Sprat, Wilkins, Boyle, Hooke, Petty, Bacon himself, continued to call in 'the aid of similitude' when they needed to explain, introduce or promote an idea, when they, in Sprat's words, wanted to make 'a strong, and a sensible impression on the *mind*' (Sprat 1959: 413).

The other point to be made is that while there is nothing in this passage that had not been adumbrated by Bacon and his followers, Sprat argued from an extreme position. His advice for the appropriate style for scientific discourse escalated into an immoderate and unreasonable attack on 'ornaments of speaking'. In Sprat's extreme, paranoiac fear of words (as seen in his statement that words will 'eat out' the 'spririt and vigour' of the design), we recognise not so much 'a reformer who brought about a change in prose style but rather ... a symptom of a general distrust of language in circles connected with the new philosophy' (B. Vickers 1987: 16). Although Sprat does not insist that words should be reduced to the level of things as, for example, in Wilkins's 'universal language' (a scheme Sprat was undoubtedly familiar with when he wrote the *History*), he does argue from that

---

[15] For Bacon's influence on Sprat's *History of the Royal Society* see Fisch and Jones 1951.

extreme position, from which the demand for an artificial language arose.

This brief outline of the seventeenth-century reform of style began with Bacon's legitimate warning against the misuse of words; it ends with those followers who, in their extreme fear of words, tried to abolish them altogether and replace them with signs representative of things. The artificial, universal language schemes were the logical conclusion to the Society's belief that a precise and unequivocal mode of expression would provide the foundation for the advancement of learning.[16] Of the various philosophical (that is, scientific) language projects that sprang up at the time, Wilkins's *Essay Towards a Real Character and a Philosophical Language* (London, 1668) was the most complete and comprehensive. In his *Epistle* Wilkins wrote that 'as *things* are better than *words*, as *real knowledge* is beyond *elegancy of speech*', his scheme will offer 'the shortest way for the attainment of real knowledge that hath yet been offered to the world'. Besides, the philosophical language will have other advantages: it will facilitate the learning of languages, help towards correcting the 'Babylonean Confusion of Tongues', and could settle religious and political controversies by eliminating linguistic misunderstandings and errors.

To this end, Wilkins set out to make 'the difficult attempt' of 'a regular *enumeration* and *description* of all things and notions'. In other words, he recorded and classified all known natural phenomena into a concrete order of genus and species. He then devised a symbol system that gave a 'distinct *Mark*' to every represented thing. Wilkins was convinced that his taxonomy of things would 'contain such a kind of *affinity* or *opposition* in their letters and sounds, as might be some way answerable to the nature of the things which they signified' (p. 21). Finally, he provided 'a *Dictionary* of the English tongue, in which shall be shown how all the words of this Language according to the various equivocal senses of them, may be sufficiently expressed by the Philosophical Tables here proposed' (pp. 1–2).

Wilkins's scheme was monumental; it was also impossible to carry out, since our traditional language, with its infinite capacity for variation, cannot be fixed upon an inflexible grid of calculation. One

---

[16] On universal language schemes see Vickery 1953, Cornelius 1965, Knowlson 1975, Fraser 1977, Salmon 1979, Clauss 1982, and Slaughter 1982.

could poke fun at the *Essay* and dismiss it as absurd, as Swift did in *Gulliver's Travels*.[17] Yet to do so would obscure the significant fact that the features that determined this scheme were identical with those underlying the Baconian scientists' call for plain prose. The reform of science and the reform of style, it must be stressed, were inextricably intertwined; both proceeded from and served the progress of knowledge. In their extreme fear of words, the universal language-schemers looked for extreme solutions. They erred because words are not 'answerable to the nature of things which they signify'; words do not, as Locke explained, stand 'for the reality of things'. Locke, defending the moderate, traditional theory of language, maintained that words are 'the signs of men's ideas', and carry the value we attach to them (Locke 1979: 405).

Wilkins's *Ecclesiastes* and Morton's *Advice*

The work in which Wilkins first pronounced his views about style and expression was *Ecclesiastes, or A Discourse concerning the Gift of Preaching* (London, 1646). *Ecclesiastes* has been recognised as a milestone not only in the history of plain preaching but in the reform of English prose.[18] Since Morton in his lecture *Advice to Candidates for the Ministry* (dated by Calamy as pre-1685) referred his students to *Ecclesiastes*, and then recapitulated many of Wilkins's ideas, it is worth looking at it in some detail.[19]

In his treatise on the art of preaching, Wilkins stressed that 'the wrestling of Scripture unto improper truths, may easily occasion the applying of them unto grosse falshoods', therefore observations

> must be laid downe in the most easie perspicuous phrase that may be, not obscured by any rhetoricall or affected expressions; for if the hearers mistake in that, all that follows will be to little purpose.
>
> (p. 12)

---

[17] In *Gulliver's Travels* in the Grand Academy of Lagado there was a project 'for entirely abolishing all Words whatsoever; and this was urged as a great Advantage in point of Health as well as Brevity ... An Expedient was therefore offered, that since Words are only Names for *Things*, it would be more convenient for all Men to carry about them, such *Things* as were necessary to express the particular Business they are to discourse on' (Swift 1972: 167).

[18] On the influence of *Ecclesiastes* on the development of English prose see Gosse 1889: 75–6, Bush 1962: 285, and Hill 1965: 130.

[19] Morton's *Advice* was printed for the first time in Calamy 1727, I: 198–210.

In the final section 'Concerning Expression' Wilkins explained his demand for a 'plain, full, wholesome, affectionate' style. A phrase, he wrote

> must be plain and naturall, not being darkned with the affectation of Scholasticall harshnesse, or Rhetoricall flourishes. Obscurity in the discourse is an argument of ignorance in the minde. The greatest learning is to be seen in the greatest plainnesse. The more clearly we understand any thing our selves, the more easily can we expound it to others. When the notion it self is good, the best way to set it off, is in the most obvious plain expression.
> (p. 72)

Turning to Morton, we find that he exactly shared Wilkins's views. He cautioned against 'oratorial Flash', which is ' a Torrent of Words' and runs 'like Water over a Mill-Wheel'. Morton taught that in the 'honest and useful Way' of writing 'more of the Christian will appear' and 'not less of the Scholar' (*Advice*, p. 207). Morton recommended the use of 'a sound Word', and 'solid, wholesome and savoury Discourses'.[20] He contrasted 'practical Holiness' with the empty notional learning of '*Schoolmen, Criticks, Theological Systems*, and *Polemicks*'. When Morton spoke of 'true Piety and Heart-Engagement', and said that only words emanating from the heart reflect the truth they seek to convey, he was echoing Wilkins's demand for language 'proceeding from the heart, and an experimentall acquaintance with those truths which we deliver' (*Ecclesiastes*, p. 73).

In 1651 Wilkins published a companion volume to *Ecclesiastes* on the art of prayer. In this discourse he advised readers to

> Beware of crude, tumultuary meditations; of idle, impertinent, wild expressions; take heed of all empty repetitions, digression, prolixity ... *Let thy words be few*; Not that brevity, or fewnesse of words is the proper excellency of Prayer ... But because those that speak little, do probably study, and ponder more upon what they say.
> (*Discourse concerning the Gift of Prayer*, p. 19)[21]

---

[20] *Advice*, pp. 205–6. By 'savoury' Morton seems to have meant a well-prepared and pleasing discourse. Wilkins similarly employed a food metaphor when he wrote: 'To deliver things in a crude confused manner, without digesting of them by previous meditation, will nauseate the hearers, and is as improper for the edification of the minde, as raw meat is for the nourishment of the body' (*Ecclesiastes*, p. 73).

[21] Wilkins's *Discourse Concerning the Gift of Prayer* is listed in the catalogue of the Defoe/Farewell libraries.

Defoe's teacher gave the same advice:

> Beware also of impertinent Repetition of Words and Sentences, which dead and flatten much the Intention of the Auditors.
>
> (*Advice*, p. 204)

That Morton believed in the phrase 'Let thy words be few' is signified by his frequently repeated saying: '*A great Book is a great Evil*' (Calamy 1727, I: 211).

While Wilkins had written: "'Tis a sign of low thoughts and designs, when a man's chief study is about the polishing of his phrase and words', Morton warned that 'the Accuracy of Speech be not more minded than the Efficacy' (*Advice*, p. 201; *Ecclesiastes*, p. 72). Wilkins had argued for a stringent correlation between 'words' and 'things': 'Our expressions should be so close, that they may not be obscure, and so plain that they may not seem vain and tedious' (*Ecclesiastes*, p. 73). Morton in his *Advice* firmly imprinted on his students' minds the importance of disposing 'Things prudently, (not Words curiously)'. And again:

> I said before, *Things* and not *Words*. Not that I advise an utter Neglect of proper and significant Expressions: But the greatest Care should be had of the Matter and Things. And if this be done, one that is a Scholar, and who ordinarily accustoms himself to speak handsomly and proper, needs not want sufficient Words well to express his Mind. *Re bene disposita, Verba ac invita sequuntur.*
>
> (p. 203)

Morton's position is 'look after the subject-matter, and words will look after themselves'; he is quoting Cato the Elder's phrase, as it had been quoted by Bacon and to the same purpose. In reflecting Wilkins's standard of prose Morton reflected the whole programme of Bacon's philosophy of things.

The direct relevance of Morton's *Advice* to the stylistic development of Defoe was noted by J. Huddlestone in 1978, but otherwise it has continued to remain 'strangely ignored by [Defoe's] critics and biographers'.[22] Exploring this evidential document more fully, we see how closely Morton's instructions followed Wilkins's and consequently the *History*'s guidelines for plain prose. Morton's stylistic advice and

---

[22] Huddlestone 1978: 37–8. And see pp. 124–5 below.

practice will help us to confront the long-standing question whether with Defoe 'the writing of English composition may be traced to the influence of the sermon-writing required in the divinity course' (Parker 1914: 71). Certainly, Wilkins's and Morton's insistence that the purpose of communication is to communicate and not to obscure or conceal facts is shared by Defoe. 'The end of Speech,' stated Defoe, 'is that men might understand one another's meaning', and he urged 'a plain and homely stile; easy, plain, and familiar language is the beauty of Speech in general, and is the excellency of all writing, on whatever subject' (*The Complete English Tradesman* (1726), p. 26). Following his predecessors, Defoe contrasted honest plainness with deceitful rhetoric: 'The plainness I profess, both in style and method, seems to me to have some suitable analogy to the subject, honesty' (*Crusoe* 3: 23). When analysing Defoe's advice, we must bear in mind Morton's *Advice*, the Baconian legacy which Morton bequeathed to his students. Morton believed that education moulded the spirit of man. 'Intellectual and Moral Habits,' he had written, 'are formed much according to the Information men meet with, especially in their younger dayes' (Morton 1693: 21–2). Defoe's teacher was convinced that his reformed practical curriculum, his up-to-date information on experimental science, together with his defence of plain prose, would form his students' habit of mind and mode of expression. Whether Morton was right in this conviction will emerge from the following discussion of Defoe's knowledge of and attitude to the New Sciences.

PART 2

Daniel Defoe

4

Daniel Defoe and the Baconian legacy

Defoe and education

One of the chief concerns of Part I of this book was to draw the reader's attention to the fact that the reform of science and the reform of education and of language were organically related. In the course of the century the three became so closely intertwined that it is virtually impossible to discuss one without reference to the other two. The question is, was Defoe aware of these related forces?

As Defoe was not a writer interested in systematically ordering his thoughts under specific headings, it is not always easy to discern his articles of faith. He was brimming over with ideas on trade, that is, with studying nature improved by the activities of men, yet he never collected these thoughts into a tractate on the subject; his continued interest in education and educational reform never found its way into a systematic treatise; equally, his recommendations on literary style and method remained scattered through his writings. However, if we collect his statements on any of these subjects, a clear and consistent pattern of thought does evolve.

For Defoe's views on education and educational reform we must turn to his advice to young tradesmen and gentlemen and to a variety of educational projects conceived over a period of thirty years. In these writings Defoe consistently attacks the outmoded curriculum for 'locking up, as I have call'd it, all science in the Greek and Latin' (*Gentleman*, p. 218). Instead of wasting time studying the classical languages, Defoe recommends that men study practical, useful subjects related to the business of life. He writes of a tutor who in order to 'rectify this great mistake' of the traditional universities, founded a 'little Accademy' where he

> taught Physicks, that is to say, Natural Phylosophy, with a system of Astronomy as a seperate science, tho' not exclusiv of the generall system of Nature; he taught also Geography and the use of the maps and globes in a seperate or distinct class: in a word, he taught his pupils all the parts of accademick learning, except Medicine and Surgery. He also had a class for History, ecclesiastic and civil. And all this he taught in English. He read his lectures upon every science in English, and gave his pupils draughts of the works of Khiel and Newton and others, translated; also he requir'd all the exercises and performances of the gentlemen, his pupils, to be made in English.
> (*Gentleman*, pp. 218–19; for a similar description of the Dissenters' education see *The Present State of the Parties in Great Britain* (1712), p. 319)

Defining the perfect, 'compleat' scholar, Defoe speaks of a student who for four and a half years attended such a private Academy. In this time:

> He run thro' a whole course of Phylosophy, he perfectly compass'd the study of Geography, the use of the maps and globes; he read all that Sir Isaac Newton, Mr. Whiston, Mr. Halley had said in English upon the nicest subjects in Astronomy and the secrets of Nature ... in those 4 yeares and half he was a mathematician, a geographer, an astronomer, a philosopher, and, in a word, a compleat scholar: and all this without the least help from the Greek or the Latin.
> (*Gentleman*, p. 207)

I am not the first to suggest that Defoe's balanced curriculum, which included natural philosophy, mathematics, history, geography, modern languages, and made English the language for instruction, mirrored his own education at Morton's Academy. The point is that by defending the Dissenters' reformed education, Defoe defended the Baconian principles upon which Morton's Academy was founded.[1]

It is in *The Complete English Tradesman* (1726) (and not, surprisingly, in his *Compleat English Gentleman*, which is replete with praise of Morton's

---

[1] Defoe was not an uncritical admirer of the Dissenters' Academies: see, for example, *Present State*, pp. 316–17; and see Leinster-Mackay 1981: 36ff.

private institution) that we find the following little vignette illustrating the inestimable value of demonstration:

> I knew a philosopher that was excellently skill'd in the noble science or study of astronomy, who told me he had some years studied for some simily, or proper allusion, to explain to his scholars the phænomenon of the sun's motion round its own axis, and could never happen upon one to his mind, 'till by accident he saw his maid *Betty* trundling her mop: surpris'd with the exactness of the motion to describe the thing he wanted, he goes into his study, calls his pupils about him, and tells them that *Betty*, who her self knew nothing of the matter, could shew them the sun revolving about itself in a more lively manner than ever he could. Accordingly *Betty* was call'd, and bad bring out her mop, when placing his scholars in a due position, opposite ... to her left side, so that they could see the end of the mop, when it whirl'd round upon her arm, they took it immediately; there was the broad headed nail in the center, which was as the body of the sun, and the thrums whisking round, flinging the water about every way by innumerable little streams, describing exactly the rays of the sun darting light from the center to the whole system.
>
> (pp. 42–3)

Morton had in fact instructed his students in the theory of the earth's motion using a very similar technique:

> Now in the Copernican Scheme *the Earth trolling itself roun[d] its own Axis (as a bowle)* in the Diurnall revolution, and withall round the Great orb ascribed to the sun by Ptolomie ... in the Annuall revolution, the Center of the Earth runing in the line of the Great orb ...
>
> (*Compendium*, Morton 1940: 64, my italics)

It was Morton's habit to accompany his lectures with diagrams explaining in the simplest terms his argument. Defoe's teacher constantly impressed upon his students' mind the persuasiveness of direct experience. 'Solid inferences', he taught, can only be made 'from well observed Experiments'. Defoe, in his turn, stresses that we must see and perceive for ourselves, we must: 'Judge of true learning by the strength of nature; reason shall be [our] guide into the study of Nature as nature shall be in the pursuit of [our] reason' (*Gentleman*, p. 188).

Defoe's complaint is that the traditional educational system produces pedagogues unfit for real life outside the walls of the university. Oxford and Cambridge turn out 'meer schollars', that is, men who 'seem to be form'd in a school on purpose to dye in a school'. (Morton's phrase had been 'the Dulheads of the universities'.) Defoe judges 'meer schollars ... a kind of mechanicks in the schools, for they deal in words and syllables as haberdashers deal in small ware'. He asks, is a student fluent in modern languages 'NO SCHOLLAR'? A man may be proficient in 'Experimental Phylosophy ... master in Geography, [and have] the situacion of the world at his fingers' ends', yet the world does not acknowledge him to be a scholar. He may be skilled in astronomy and history but 'he is NO SCHOLLAR'. Defoe's next point is especially revealing. Recalling the circumstances in which he had a few years earlier gathered the details for the *Tour*, he writes of one who had made a survey of England and in the process had become

> For his own country ... a walking map; he has travell'd thro' the whole island, and thro' most parts of it severall times over; he has made some of the most criticall remarks of severall parts of it, so that he could not be charg'd, when he went abroad, to have known much of other countryes and nothing of his own as is the just scandal of most English travellers: and yet this man forsooth is NO SCHOLLAR.
>
> (*Gentleman*, pp. 200–1)[2]

While it would be wrong to ignore the personal, self-defensive note in this invective, it should not blind us to the fact that Defoe is advocating the kind of practically orientated curriculum that had been proposed by the Baconian Puritan reformers of education. When Defoe wars against 'the dead learning of tongues', he echoes their earlier criticism of 'the grammatical Tyranny of teaching Tongues'; when he demands that the study of Latin and Greek be reduced and made secondary to the study of sciences, he echoes what Hartlib and his men had recommended in their reformed schools (see p. 21 and pp. 36–7 above).

The Baconian educational reformers had placed a marked emphasis on things, not words, on useful knowledge, not unprofitable notions. To recall some statements made in the middle of the seventeenth

---

[2] See *The Great Law of Subordination consider'd* (1724), pp. 46–7, where Defoe describes the way in which he had originally collected his information for the *Tour*; the quotation is given in full in chapter 8, p. 160.

The same distinction between 'the meer scholar' and 'the man of polite learning' is made in *Applebee's Journal*, in Lee 1869, III: 435–7 and in *Present State*, pp. 316–17.

century: Milton in his pamphlet 'Of Education' (dedicated to Hartlib in 1644) had written *'that language is but the instrument conveying to us things useful to be known.* And though a linguist should pride himself to have all the tongues ... if he have not studied the solid things in them as well as the words and lexicons he were nothing so much to be esteemed a learned man as any yeoman or tradesman competently wise in his mother dialect only' (Milton 1973: 48, my italics). Hartlib, Dury, Petty convinced their age 'that the teaching of *words* is no further usefull than the [teaching] of *things*' (my italics). Typical in this respect was also John Aubrey who in *An Idea of Education of Young Gentlemen* (published for the first time in 1972) stated that 'Nature is the best Guide, & the best Paterne: 'tis better to copie nature than Bookes: as the best Painters imitate nature, not copies' and he cited Hobbes, Petty, Wren and Hooke as having made similar statements (M. Hunter 1975: 40 and 235). Towards the end of the century Aubrey's friend, John Locke, published *Some Thoughts Concerning Education* (London, 1693). Locke, a devoted (albeit unacknowledged) disciple of Bacon, recommended that the learning of languages should be joined with as much other real knowledge as possible and 'beginning still with that which lies most obvious to the senses; such as is the knowledge of minerals, plants, and animals, and particularly timber and fruit-trees ... But more especially geography, astronomy, and anatomy' (Locke 1922: 138). And in the light of the discussion in chapter 3, we recall Morton's stylistic instructions to his pupils. In his *Advice* Morton had insisted upon the importance of disposing 'Things prudently (not Words curiously)'; Defoe's teacher urged his students to look after '*Things* and not *Words*' (*Advice*, in Calamy 1727, I: 202–3).[3] The Baconian educational reformers used the word–thing opposition to express their stress on a realistic approach to learning. The emphasis was on 'real' learning, from '*realis*', the late Latin adjective from '*res*' which in the seventeenth century had gradually changed from 'subject matter' to 'things' in the sense of material objects (cf. Howell 1946: 131, and see n. 13 p. 43 above).

This is the tradition into which Defoe's attitude to education and educational reform fits. In the company of Bacon, Milton, Hartlib, Petty, Aubrey, Morton and Locke, Defoe declares that 'the knowledge

---

[3] See pp. 43 and 50–1 above and p. 125 below.

of things, not words, make a schollar'. Clarifying his stress on the knowledge of things, he goes on to compare the true scholar with Solomon, Bacon's favourite searcher into nature. Defoe acknowledges that man to be a 'compleat scholar' who 'according to Solomon ... seeks for Knowledge as for silver and ... searches for her as for hid treasure' (*Gentleman*, pp. 212 and 37). A sceptic might ask whether Defoe's references to the Baconian prototype, Solomon, and the word–thing dichotomy do in fact reveal a deeper understanding of the relation between the acquisition of knowledge and language. But there is other evidence to show that Defoe was aware of this critical issue of the time. Echoing in particular Milton's concept of language being 'the instrument conveying to us things useful to be known', he writes in *Mere Nature Delineated* (1726):

> *Words* are to us, the Medium of Thought; we cannot conceive of *Things*, but by their *Names* ... we cannot muse, contrive, imagine, design, resolve, or reject; nay, we cannot love or hate, but in acting upon those Passions in the very Form of *Words*; nay, if we dream 'tis in *Words*, we speak every thing to ourselves, and we know not how to think, or act, or intend to act, but in the Form of *Words*.
>
> (pp. 38–9, my italics)[4]

[4] For Defoe's other references to the word–thing dichotomy see *Projects* (p. 130). Defending here the idea that a direct correlation between words and things eliminates ambiguities, Defoe writes: 'But there is a direct signification of words, or a cadence in expression, which we call speaking sense; ... and there is a superfluous crowding in of insignificant *words* more than are needful to express the *thing* intended, and this is impertinence; and that again, carried to an extreme, is ridiculous' (my italics). Shortly afterwards the word–thing (or word–works) opposition is used in *The True-Born Englishman* (1701). Defoe makes Britannia exclaim:

> *Then seek no Phrase his Titles to conceal,*
> *And hide with Words what Actions must reveal.*
> *No Parallel from Hebrew Stories take,*
> *Of God-like Kings my Similies to make:*
> *No borrow'd Names conceal my living Theam;*
> *But Names and Things directly I proclaim.*
>
> (ll. 919–24)

Again the word–thing correspondence vouches for honesty and a true understanding of the matter in hand. Considering in an issue of the *Review* the possibility of curing the English of their 'perverseness' in politics by describing 'the Thing it self', Defoe observes:

> For my part, I can't pretend to the Cure, I'll leave that to the Learned; but I may talk a little to you of the Disease, and draw the Picture of the Thing for you. Perhaps if we could but artfully enough describe the Ugliness and Deformity of the Thing it self, we might stamp some Aversions in Mens Minds to it – By the Doctrine of Idea's it is allow'd, That to Describe a Thing, Ugly, Horrid and Deform'd, is the best way to get Abhorrence in the Minds of the People – and this was the Method of the great Men in

In his advice on education, Defoe's stress on the study of things reveals his unmistakable familiarity with this movement of ideas. His views on education reflect the typical Baconian attitude towards 'real', practical learning, the importance of which cannot be too strongly emphasised. Defoe's concern with the facts of experience profoundly influenced his concept of the world of nature and of man, it gave him a habit of mind that permeated everything he thought and wrote.

In spite of the fact that Defoe repeatedly makes his demand for life-related knowledge within the context of his praise of Morton's Academy, no detailed study that takes account of both Morton's and Defoe's intellectual indebtedness has ever been made. Indeed, as I have tried to show, Morton's intimate familiarity with Baconian experimental science has so far gone unnoticed. It was in Morton's lectures, full of references to Baconian scientific inventions and discoveries that

> the East, in the Ages of Hieroglyphicks, when Things were more accurately Describ'd by Emblems and Figures than Words; and even our Saviour himself took this Method of Introducing the Knowledge of himself into the World, (*viz.*) By Parables and Similitudes.
>
> (VII: 25)

When M. E. Novak commented on this passage he wrote: 'Defoe's indebtedness ... is directly to John Locke.' While Defoe might 'have derived his idea of "hieroglyphics" and the phrase "the thing itself", from Bacon or any number of writers', it is the reference to the 'Doctrine of Ideas' that gives the game away and reveals Defoe as a disciple of Locke (1964: 661–2). I think it is important to remind ourselves at this point that Locke himself was a follower of Bacon. Such fundamental ideas in *An Essay concerning Human Understanding* (London, 1690), as that words 'came to be made use of by men as the signs of their ideas; not by any natural connexion ... but by a voluntary imposition, whereby such a word is made arbitrarily the mark of such an idea', can be traced directly to Bacon. Secondly, if we compare the passage from the *Review* with Bacon's original proposal that as 'Real Characters' and 'Hieroglyphics' 'represent ... things and notions' they might offer a more direct mode of communication, we find that there is a striking similarity between the two statements. Bacon had written: 'The use of Hieroglyphics is very old, and held in a kind of reverence ... it is plain that Hieroglyphics ... have always some similitude to the thing signified, and are a kind of emblems.' Compare this to Defoe's sentence: '... in the Ages of Hieroglyphicks, when Things were more accurately Describ'd by Emblems and Figures than Words'. Defoe's speculation that by describing 'the Thing it self, we might stamp some Aversions in Mens Minds to it' may have been triggered off by Bacon's observation that 'we are handling here the currency (so to speak) of things intellectual, and it is not amiss to know that as moneys may be made of other material besides gold and silver so other Notes of Things may be coined besides words and letters' (Bacon 1857, IV: 439–40). To be sure, these ideas changed and developed in the course of the century (Singer 1989). The point I would like to make is that in Defoe's reference to hieroglyphics we have one further instance where the author reveals his knowledge of Bacon. I do not claim that Bacon was the only influence on Defoe's idea of hieroglyphics but that it was an important one and has been strangely neglected.

Defoe first became acquainted with these ideas. Here he was introduced to the idea that the knowledge of things, not words, made a scholar. Here he learnt that knowledge must be shared and that this main Baconian ideal was defeated by the use of Latin and Greek. Almost fifty years after his experience at the Dissenters' Academy, Defoe writes that 'Science being a publick blessing to mankind ought to be extended and made as difusiv as possible, and should, as the Scripture sayes of sacred knowledge, spread over the whole earth, as the waters cover the sea' (*Gentleman*, pp. 197–8).[5] At Morton's reformed Academy Defoe realised that the study of a wide range of subjects, combining 'real' learning with the liberal arts, was the foundation for a truly 'compleat' education. As we have seen, Morton, together with the first-generation Baconians, argued for a balanced curriculum; Defoe never forgot this advice (see pp. 36–8 above).

Defoe's teacher believed that early education had an especially powerful effect on the mind of man. 'Intellectual and Moral Habits', he had written, in a tractate entitled *The Spirit of Man*, 'are formed much according to the Information men meet with, especially in their younger dayes' (Morton 1693: 21–2). This is the view that Defoe defended from his first important published work to the last. At a tender age the mind is impressionable, the soul is

> as a Lump of soft Wax, which is always ready to receive any Impression; but if harden'd, grow callous, and stubborn, and like what we call Sealing-Wax, obstinately refuse the Impression of the Seal, unless melted, and reduced by the Force of Fire; that is to say, Unless moulded and temper'd to Instruction, by Violence, Length of Time, and abundance of Difficulty.
>
> (*Mere Nature Delineated*, pp. 60–1)[6]

Using a very different metaphor to enlarge upon the benefits of education, Defoe had written thirty years earlier in connection with his project for 'An Academy for Women':

> The soul is placed in the body like a rough diamond, and must be

---

[5] And compare Defoe's view that it is our moral and religious duty to share knowledge, with Morton's insistence 'to have Knowledge encreas'd, and not only confin'd to the Clergy or Learned Professions, but extended or diffus'd as much as might be, to the People in general' (*Advice*, in Calamy 1727, I: 192).

[6] Locke had compared the developing mind to 'white paper, or wax, to be moulded and fashioned as one pleases' (Locke 1922: 179).

polished, or the lustre of it will never appear: and it is manifest that as the rational soul distinguishes us from brutes, so education carries on the distinction and makes some less brutish than others. This is too evident to need any demonstration.

(*Projects*, p. 145 and cf. *History of Arts and Sciences*, p. 1)

Education moulds the human soul. The 'gifts of nature', Defoe writes, are 'improv'd at school: those rough diamonds are polished by the schools and by the help of books and instruccion'; the 'bright genius must be made brighter by art' (*Gentleman*, pp. 55, 109). Both in his analogy and in his desire for clear, solid and 'real' knowledge, Defoe reflects the fundamental tenets of the Royal Society whose 'principles,' Boyle had declared, 'ought to be like diamonds, as well very clear as perfectly solid'.

Perhaps no other aspect of Defoe's writing gives us as clear an insight into his habit of mind as his advice on gathering and using knowledge. A corollary of his practical approach to learning is his interest in educational projects. 'We always thought', he writes in an issue of the *Review*, that 'Women had the quickest and justest Notions of things at first sight, *'tho we have unjustly rob'd them of the Judgment, by denying them early Instruction.*'[7] The curriculum that he suggests for women's colleges includes not only music and dancing, modern languages, polite conversation but 'especially history': women should learn 'to read as to make them understand the world, and be able to know and judge of things when they hear of them' (*Projects*, p. 148). The same practical approach that we noted in Defoe's advice for gentlemen and tradesmen is to be found in his advice to women. His educational projects, like his other projects for the reform of trade, society or religion are written for the 'advancing of the Interest of the Nation', or, as he puts it in one of his last works, with the 'Publick Good in View' (*Augusta Triumphans*, 1728).

In 1728, three years before his death, Defoe writes:

> I have but a short Time to live, nor would I waste my remaining Thread of Life in Vain, but having often lamented sundry Publick Abuses, and many Schemes having occur'd to my Fancy, which to me carried an Air of Benefit; I was resolv'd to commit them to Paper before my Departure ...
>
> (*Augusta Triumphans*, p. 4)

[7] *Review*, I: 156.

So, as a testimony of his 'good Will to [his] Fellow Creatures', Defoe puts forward the proposal for a university in London. *Augusta Triumphans* consists of a number of projects for the protection of battered wives, for the prevention of women being sent by their husbands to mad-houses, for the prevention of street-robberies, etc. These social schemes are preceded by the project for a university in the metropolis: 'of all my Reflections, none was more constantly my Companion than a deep Sorrow for the present decay of Learning among us and the manifest Corruption of Education' (ibid., p. 4). Since, as we have seen, Defoe frequently criticises the traditional universities, his plan for a college in London is the logical extension of his criticism.

It must, however, be pointed out that his idea for a university in London was not new. By the middle of the seventeenth century several schemes for universities outside Oxford and Cambridge had been proposed by the Baconian Puritan reformers. Petty had suggested in his *Advice of W.P.* of 1648 that a university be founded in London, and in 1657 a college 'for all the sciences and literature' was actually instituted in Durham; this college, with Hartlib on its committee, existed until 1660.[8] Again, the idea that learning was not the monopoly of the male sex, was part of the Baconian Puritan commitment to educational reform.[9] Hartlib, with the help of his protégés, provided 'Parliament with plans for a complete system of education, embracing research, teacher training, inspection, schools and workhouses for all social classes and both sexes'.[10]

While Defoe refers to Gresham College, the birth-place of the Royal Society (it is 'a kind of University ... where Professors in all Sciences are maintained and obliged to read Lectures every Day, or at least as often as demanded', *Augusta Triumphans*, p. 7), he does not make reference to any of these other schemes, either proposed or initiated. This does not necessarily mean that he was unaware of them; Defoe was not an author who paraded the total range of his knowledge. He

---

[8] Hill 1965: 109, 124. And see the Appendixes I and II in C. Webster 1975: 523–32. For Defoe's proposal for a College of Trades, the '*QUEEN ANNE'S COLLEGE OF INDUSTRY*', of London, see *Proposals for Imploying the Poor. In and about the City of London* (1713), pp. 9ff.

[9] John Dury's wife wrote a brief sketch entitled 'Of the Education of Girls' which is given in full in Turnbull 1947: 120–1. For proposals for an Academy in London see Turnbull 1952–3: 104 n.23 and Syfret 1947–8: 112–13.

[10] C. Webster 1975: 210.

may or he may not have known of these projects; certainly his inspiration for a metropolitan university flowed from the same source which served his enduring enthusiasm for useful learning for the good of mankind.

Defoe and the New Sciences

Defoe best defines the purpose and limitations of experimental science in *The Storm* (1704), a work which records a series of eye-witness accounts of the great tempest of 26–27 November 1703. Aware of his responsibility as a writer, Defoe promises in the Preface 'to be careful of his words, that nothing pass from him but with an especial sanction of truth' (p. 251). 'Though the undertaking be very difficult amongst such infinite variety of circumstances' and though the stories to be related verge on the incredible, Defoe vouches 'to keep exactly within the bounds of truth'. Guarding against any errors, he writes, 'if', despite his efforts, 'the least mistake happen, it shall not be mine'. Finally, apologising for 'the meanness of style', he comments that he does not aim at eloquence but rather that his thoughts should be 'dressed in the desirable, though unfashionable garb of truth' (pp. 252–8). The question is, is Defoe's apparent scrupulous regard for truth mere convention, and are these statements guaranteeing reliability, the kind of statements that we would expect to find in the works of a hack-writer turned journalist; or may we, on the other hand, recognise in Defoe's factual report a genre that had evolved from the activities of the Royal Society?

That the Royal Society's guiding ideals encompassed objectivity, accuracy, scepticism is common knowledge. What is not always realised is that not content with adhering to these ideals, the Fellows felt the need to explicitly state that they recorded nothing but '*severe, full and punctual Truth*'.[11] Their histories of nature were, as Sprat had declared, 'faithful *Records* of all the Works of *Nature* or *Art*'. In order to fulfil this ambitious aim of recording all aspects of human knowledge, the members of the Society found it necessary to enlist outside help. They turned to the 'unlearned', that is, to 'Seamen, Travellers, Tradesmen,

---

[11] *Philosophical Transactions*, General Preface to vol. XI (1676–7), p. 552.

and Merchants' and asked them to contribute with their personal experience to the Society's repository of facts (Sprat 1959: 155). In the service of the Society these '*experienc'd Men*' adopted the experimental scientists' standards and meticulously observed and recorded information about the physical universe. Moreover, when they sent in their reports they repeatedly stressed that their accounts were truthful, and that they had recorded only what they had witnessed with their own eyes.[12]

The idea that a comprehensive and exact view demanded the assistance of many observers is at the centre of *The Storm*. Describing how he collected his material, Defoe writes: 'we have endeavoured to furnish ourselves with the most authentic accounts we could from all parts of the nation'. It seems that for his accurate description of the storm of 1703 Defoe issued a nation-wide call for aid, and that his call was answered by 'a great many worthy gentlemen [who] have contributed their assistance in various, and some very exact relations and curious remarks' (p. 298).

In this Defoe imitated the Baconian scientists' method of procedure. In December 1696, less than a decade before the publication of *The Storm*, Edward Lhwyd had in identical fashion circulated a letter in Wales asking 'the Gentry and Clergy' to contribute with their firsthand experience to his projected Natural History of Wales. Lhwyd's 'Parochial Queries' had been posted with the sanction of John Wallis, Martin Lister, John Ray and Edward Bernard. Explaining its purpose Lhwyd wrote:

> Having Publish'd some Proposals towards a Survey of *Wales* ... I thought it necessary for the easier and more effectual Performance of so tedious a Task, to Print the following Queries; having good Grounds to hope the Gentry and Clergy ... will also readily contribute their Assistance, as to Information; and the Use of their Manuscripts, Coyns, and other Monuments of Antiquity: The Design being so extraordinary difficult without such Helps, and so easily improvable thereby ... My Request therefore to such as are desirous of Promoting the Work, is, That after each Query, they would please to write on the blank Paper ... their Reports ... distinguishing always betwixt Matter of Fact, Conjecture, and Tradition.

---

[12] For a fuller discussion of the relationship between the Royal Society and the Restoration traveller see chapters 7 and 8.

Lhwyd ends by expressing the characteristic Baconian view that for his 'compleat Account' everything should be noted down, 'seeing that what we sometimes judge insignificant, may afterwards upon some Application unthought of, appear very useful' (R. Ellis 1906–8: 17–18). Again, John Aubrey recorded that the Society 'have been pleased to lay their Commands upon me to keepe a Correspondence with my numerous company of ingeniose Virtuosi in severall Counties ... for *things naturall* or *artificiall* or any thing remarkqueble in phylosophy, or mathem[atics]' (M. Hunter 1975: 64).

Similarly Defoe, by furnishing himself with the most authentic reports 'from the country', hopes to offer a complete and exact view of the storm. That he had the experimentalists' method in mind is shown by the fact that he not only mentions Bacon, Boyle and Harvey by name but uses excerpts from the *Philosophical Transactions* as proof of the reliability of his report. Of the replies that Defoe prints, many explicitly state that 'the account is authentick', that it is 'a very exact and faithful account' or 'a true and exact account', and so on. If the letter itself does not carry a warranty of truth, Defoe will introduce it with a comment that the 'account [is] very authentic' and from 'a gentleman whose credit we cannot dispute' (p. 308), or from a person 'of undoubted credit and reputation' (p. 310). Defoe's aim is obvious enough: by citing reliable eye-witness accounts he seeks to give authenticity to his report, assuring his readers that 'the truth of [*The Storm*] may be depended upon'. It is possible that Defoe invented these letters. If so, his adherence to the ideals of the observational method is not diminished. Whether he has invited or invented the eye-witness accounts, either way he testifies to his belief in personal observation and personal experience.

Whatever our feelings may be about the quoted private letters, the authenticity of the excerpts from the *Philosophical Transactions* with which Defoe supplements his report cannot be doubted. Two of these are especially interesting. Defoe quotes 'Part of a Letter from Mr. Anthony van Lauwenhoek [*sic*], F.R.S.'. It appears that when the storm arose Leeuwenhoek had only one thought and that was 'to make my observations' and experiment whether the water 'dasht against the glass-windows' was rain or sea-water. Leeuwenhoek's letter is less a report about the storm than about his experiment. The fact that Defoe should have chosen to print this letter is evidence that he, too, was curious about how far 'the sprye of the sea ... might, by the impetuosity of the winds be carried' (pp. 349–51).

Defoe also quotes from 'the ingenious antiquary, Mr. Thoresby, of Leeds'. While the full significance of Defoe's knowledge of Thoresby's combined scientific and antiquarian activities will become clear later, it may be said here that Thoresby was a devoted follower of Bacon. In 1695 he was one of several Fellows who contributed with their expert knowledge to Gibson's revision of Camden's *Britannia*, that is, to the work that was constantly within reach while Defoe wrote the *Tour*.[13] Although Defoe liberally uses the New Scientists' 'Additions' (especially for Northern England, the section for which Thoresby was responsible), he does so without acknowledging his source. The discovery now that Defoe was perfectly aware of Thoresby's up-to-date and accurate scientific research puts his silent borrowing in the *Tour* in a new light. With Defoe the reader just can never assume that the author's knowledge is spelled out in the text. By citing Thoresby and Leeuwenhoek Defoe does what the Fellows of the Society regularly did in their reports. He appeals to other writers as witnesses or 'certificates to attest matters of fact'. With the experimentalists, Defoe hopes that the cited 'testimonies will as well be illustrated by mine, as mine by their's, and that all of them may contribute to [the readers'] better information' (Boyle 1744, I: 201).

For Defoe, as for the Fellows, the amassing of circumstantial evidence serves the larger aim of determining fundamental laws governing the universe. In the words of the sub-title of chapter 1, the purpose of *The Storm* is the discovery of 'the natural causes and original of winds'. Defining the experimental philosopher, Defoe writes that it is not enough for man to study nature but he must find out where everything is placed, 'and why there; and what their business, what their influences, their functions, and the end of their being'. His advice is: '*Search into the method nature proceeds upon* in the performing the office appointed, ... *search the steps she takes*, the tools she works by; and, in short, know all that the God of nature has permitted to be capable of *demonstration*' (p. 262, my italics). Even if Defoe had not referred to

---

[13] See p. 161 below.

Bacon just before and after this passage, it would be clear that he expresses his allegiance to Baconian science.

To the question whether science and Christian faith are compatible, Defoe answers:

> Not but that a philosopher may be a Christian, and some of the best of the latter have been the best of the former, as Vossius, Mr. Boyle, Sir Walter Raleigh, Lord Verulam, Dr. Harvey, and others ...
>
> And it seems a just authority for our search, that some things are so placed in nature by a chain of causes and effects, that upon a diligent search we may find out what we look for: to search after what God has in his sovereignty thought fit to conceal, may be criminal, and doubtless is so; and the fruitlessness of the inquiry is generally part of the punishment to a vain curiosity: but to search after what our maker has not hid, only covered with a thin veil of natural obscurity, and which upon our search is plain to be read, seems to be justified by the very nature of the thing, and the possibility of the demonstration is an argument to prove the lawfulness of the inquiry.
>
> (pp. 260–1, 262–3, and cf. Bacon 1857, III: 222, IV: 88–9 and 257)

With Bacon and his followers, Defoe believes that 'being addicted to Experimental Philosophy, a Man is rather assisted than Indisposed, to be a good Christian' (see pp. 14–15 above).[14] In his later writings he repeatedly returned to the difficult question of the reconcilability of science and religion. In a work published at the same time as *The Storm* he concluded for the time being that 'when Men Pore upon these Sacred Mysteries of Religion with the Mathematical Engines of Reason, they make such a incoherent stuff of it, as would make one pity them'.[15]

One year after *The Storm* Defoe published *The Consolidator* (1705). Although a deeply imaginative work, *The Consolidator* lacks coherence and focus. It is written in the long literary tradition of 'moon voyages', and is also an allegory of the religious and political quarrels of the

---

[14] Morton had taught that 'it is not impertinent for men to be inquisitive into the mystery of *nature natured* (the Creation) which is the work of *nature naturing* (the creator) because it makes for the Glory of God; and our own Good: Yea it is his command that we should meditate upon all his works both of Creation and Providence ... So that you See 'tis natural Theology that men should be industrious in Natural Phylosophy' (*Compendium*, Morton 1940: 4).

[15] *An Enquiry into the Case of Mr. Asgil's General Translation* (1703), p. 8, and see Merrett 1980: 9–29. For a fuller discussion of the relationship between science and religion see pp. 112–20 below.

time.[16] To us, it is significant for its remarkably detailed commentary on the activities of the members of the Royal Society.

Writing of a Royal Society in the 'moon-world', Defoe observes that 'all our philosophers are fools, and their transactions a parcel of empty stuff, [compared] to the experiments of the Royal Societies in this country'. He goes on to describe a lunar 'learned tract of winds', experiments made with 'glasses of hog's eyes that see the wind', and calculations that can prophesy 'how many storms there shall happen to any period of time, and when' (p. 272). Defoe pretends to be inventing these experiments, when in fact he describes some of the activities carried out by the Fellows of the Royal Society of London. As we have seen, Hooke, Boyle, Wilkins, Wren and Locke were engaged in making Histories of the Weather (see p. 30 above and pp. 106–10 below) and Defoe's teacher, Charles Morton, advised in his science lectures that 'a Register [be] kept of all Changes of Weather; that So a probable conjecture may be made of what will be, by a comparing of what has been already; which thing would be of Excellent Political, and ... Œconomical Use'.[17]

The narrator of *The Consolidator* reports on studies on the flux and reflux of the sea, the theories of tides, and the invention of the telescope and microscope. We are told of a philosopher who explored 'the secrets of nature', and studied 'how sensation is conveyed to and from the brain, why respiration preserves life, and how locomotion is directed to, as well as performed by the parts' – anyone familiar with the beginning of modern science will recognise in this list a reflection of the experiments of the New Scientists. Defoe's comment that 'they very often find one thing when they are looking for another', precisely describes the accepted method of the experimental scientists (p. 273).

Of special interest is his reference to a 'speaking-trumpet ... to convey sound' (p. 281). Among the riches of Solomon's House in *New Atlantis* were the 'sound-houses' which provided the 'means to convey sounds in trunks [that is, tubes] and pipes, in strange lines and distances' (Bacon 1857, III: 162–3). Wilkins had built at Wadham a

---

[16] See Nicolson and Mohler 1937: 420ff.

[17] Morton had discussed the cause of storms in his system 'Of Aiery Meteors', as we have seen, a chapter directly inspired by Bacon's suggestion for making histories of storms (*Compendium*, Morton 1940: 28–9, 94–9, and see p. 41 above).

'hollow Statue which gave a Voice, & utterd words, by a long & conceald pipe which went to its mouth, whilst one spake thro it, at a good distance' (see p. 35 above). Charles Morton, too, had explored the nature of sound with the help of a speaking trumpet. An amusing anecdote is attached to Morton's experiment. It is reported that Samuel Wesley was shocked when his schoolmates tried out Master Morton's speaking trumpet and bellowed insults at the local Anglican clergyman.[18] More seriously, Defoe's apparent flight of the imagination once again is based on the work of the Baconian scientists.

Defoe writes of a lunar 'great searcher into nature' who has invented 'various engines and curious contrivances to go to and from his own native country, the moon'. And he goes on:

> All our mechanic motions of Bishop Wilkins, or the artificial wings of the learned Spaniard,[19] who could have taught God Almighty how to have mended the creation, are fools to this gentleman.
>
> (p. 280)

Wilkins – as Defoe obviously is aware – had made a significant contribution to the tradition of moon voyages. In his *Discourse* (London, 1640) Wilkins had conjectured: 'I do seriously, and upon good grounds affirm it possible to make a flying-chariot; in which a man may sit, and give such a motion unto it, as shall convey him through the air' (Wilkins, 1640: 238). Most likely inspired by Wilkins, Morton had suggested that 'Artificiall wings might be made proportionable to the body of a man' and that 'as oars have been Substituted for finns to [enable] men to pass on water, So after ages may find out a Substitute for Wings so applied to an Engine, that may row us in the Air, as a boat doth in water' (*Compendium*, p. 191). Influenced by these speculations, Defoe writes of an 'engine formed in the shape of a chariot, on the backs of two vast bodies with extended wings, which spread about fifty yards in breadth, composed of feathers so nicely put together that no air could pass' (p. 281). Wilkins had imagined that such an engine 'might be made large enough to carry divers men at the same time,

---

[18] Morton's history 'Of Hearing' made extensive use of Sir Samuel Morland's *Account of the Speaking Trumpet* (London, 1672) printed in the *Philosophical Transactions*, see *Compendium*, Morton 1940: 165n.

[19] The reference is to Francis Godwin's *The Man in the Moone: Or, A Discourse of a Voyage thither*, by Domingo Gonsales (London, 1638).

together with food for their *viaticum*, and commodities for traffic', while Defoe's 'famous Engine [can carry] a whole Nation up to the World in the Moon'.

The experimental scientists saw in the 'flying engines' an invention providing both 'pleasure and profit': 'For besides the strange discoveries that it might occasion in this other world, it would be also of inconceivable advantage for travelling, above any other conveyance that is now in use ... In brief, do but consider the pleasure and profit, of those later discoveries in America, and we must needs conclude this to be inconceivably beyond it' (Wilkins 1640: 239, 242). Defoe's narrator reflects this characteristic attitude of the Baconians when he describes the 'flying-chariot' as being above all the inventions the most 'pleasant or profitable' (p. 281).

When it comes to explaining the principles upon which the 'flying engine' operates, Defoe has relatively little to say:

> when this engine, by help of these artificial wings, has raised itself up to a certain height, the wings are as useful to keep it from falling into the moon as they were before to raise it and keep it from falling back into this region again.
>
> This may happen from an alteration of centres; and gravity having passed a certain line, the equipoise changes its tendency; the magnetic quality being beyond it, it inclines of course, and pursues a centre, which it finds in the Lunar world, and lands us safe upon the surface.
>
> (p. 291)

Defoe neither criticises nor satirises the possibility of human flight, instead he optimistically accepts the idea and then uses it for political satire. Defoe's position becomes clearer when we compare the 'Consolidator' with Swift's 'Flying Island' in *Gulliver's Travels*.[20] Swift conceives of the ingenious but impossible idea of a 'floating island' which operates upon the principles of terrestrial magnetism. The focus of Swift's satire is the experimental scientists' belief that art could conquer nature. Swift lets the flying island get out of control so that 'the Officers ... found the Descent much speedier than usual, and by turning the Loadstone could not without great Difficulty keep it in a

---

[20] On the possible connections between *The Consolidator* and *Gulliver's Travels* see Nicolson and Mohler 1937: 425ff and Ross 1941.

firm position, but found the Island inclining to fall'. In the end the flying islanders find themselves threatened by that which should have been their greatest protection (Swift 1972: 155–6). While Swift satirises the inventions of the Royal Society, Defoe sets out to satirise English politics; Swift's satire hit the mark, Defoe's misfired. However, as a record of Defoe's imaginative use of the experimental scientists' interests and achievements, *The Consolidator* is significant.

Although *The Storm* and *The Consolidator* were composed relatively early in Defoe's career as a writer, he was over forty and all the essential features that were to determine his later writings – his life-long absorption in useful knowledge on subjects like education, trade, economic and social reform, and his corresponding concern with a factual, objective and reliable rendering of his thoughts – are evident in these works. Recognising science as an instrument with which to unlock the secrets of nature, Defoe expresses in these works his confidence in the new approach. In a hymn to science written at this time he exclaims:

> *Hail Science*, Natures *second Eye*,
> *Begot on* Reason *by Philosophy*,
> *Man's Telescope* to all that's *Deep* and *High*;
> > *What Infinites* dost thou pursue!
> The *Tangled Skeins of Nature* how undo!
> > Pierce all her darkest Clouds, *her Knots untye*,
> And leave her naked to the wandring Eye.
>
> (*Caledonia*, p. 8)

Recalling the fundamental Baconian notion that man must 'dig further and further into the mine of natural knowledge' and bring to light new facts and ideas,[21] twenty years later Defoe writes in his *History of Arts and Sciences*:

> I cannot believe that God ever design'd the Riches *of* the World to be useless *to* the World; that the Gold, the Silver, the Diamonds, and other Species of such Immense Worth and Value, was ever created in the Bowels of the Mountains, and the most hidden Parts of the World to lye burried there, and remain unprofitable.
>
> (pp. 2, 6, 266 and cf. *History of Trade*, I: 26, 40)

---

[21] Cf. Bacon 1857, III: 219, and p. 503 where Bacon speaks of knowledge being 'digged out of the hard mine of history and experience'.

Expressing another Baconian idea, Defoe encourages mankind to 'unlock the treasure-house of nature'. He asks:

> Why shou'd we not enter upon these Works again, that the Inexhausted Treasure being farther search'd into, and the discovering the bottom of them re-assumed, all that Nature had hid, and the Almighty has reserv'd for us, may be found out, and the World made as Rich, and as Learned, in all necessary Knowledge, as they were intended to be at first? and that as their Maker has made a plentiful Provision for them, they may let him know how ready they are to accept of, and improve it?
>
> (*History of Arts and Sciences*, p. 7)

Like Bacon, Defoe rejects the traditional reverence of antiquity. He insists that the ancients have not attained the sum of human achievement but, 'nobly led us by the Hand to the very Door, where what remains is to be found' (p. iv). Bacon's argument had been that the ancients had been content to read the Book of Nature as it lay before them, while the experimental scientists should try to find out new ways of adapting and improving the gifts of nature. Defoe observes that past ages 'have open'd [Nature's] Book, and read far in it, but she shows us many Leaves, not yet turn'd over; and assures us, she has reserv'd sufficient to encourage, and to reward our future Enquiries' (p. iv). Calling his age to produce further inventions and discoveries, he writes: 'Thus the World is daily encreasing, particularly in *experimental Knowledge*; and let no Man flatter the Age with pretending we are arriv'd to a perfection of Discoveries' (p. 240, my italics).

Although by the beginning of the eighteenth century the battle against the ancients may have been won, Defoe finds it necessary to re-engage in the debate and to stress the superiority of the moderns. Our progenitors 'were ignorant of Places, so of Things also', but now it is 'as if all Nature was newly laid open'. 'Infinite Experiments' were made 'by the *Boyls* [sic] *and Newtons*' of the present age, and 'all modern Knowledge seems to have built upon their first Experiments and to stand upon their Shoulders' (p. 266). One could argue that such questions as the relationship between the 'ancients' and 'moderns', established authority and intellectual independence, words and things, were, and always will be part of the thinking world.[22] Defoe's engage-

---

[22] Of the eighteenth-century writers who resuscitated the war against the ancients, Oliver

ment in the debate is, however, not general and undefined; he lays hold of these dichotomies in order to defend his position as a 'modern', and he does so by using the methods and terminology characteristic of the experimental scientist. He asks:

> What was the World before? ... Where were the Men that arriv'd to Characters, to Fame, and to Distinction, by Trade, by the Mathematicks, by the Knowledge of natural or experimental Philosophy? Where was [sic] the Sir *Walter Raleighs*, the *Verulams*, the *Boyls*, or *Newtons* of those Ages? Nature being not enquir'd into, discover'd none of her Secrets to them, they neither knew, [n]or sought to know, what now is the Fountain of all human Knowledge, and the great Mistery for the Wisest Men to search into, I mean *Nature*.
>
> (pp. 238–9)

A clearer statement of Defoe's appreciation of the achievements of the New Science cannot be found. Before these great searchers into nature had applied their new methods, he declares, the world had

> *Philosophy* without Experiment.
> *Mathematicks* without Instruments.
> *Geography* without Scale.
> *Astronomy* without Demonstration ...
>
> (pp. 233–4)

Writing of Russia before Peter the Great had introduced his economic and social reforms, Defoe similarly comments:

> Their best surgeons knew nothing of anatomy; their best astronomers knew nothing of ecclypses; they had not a skeleton in the whole empire, except what might be natural in their graves; their geographers had not a globe; their seamen not a compass (by the way they

---

Goldsmith helps best to illuminate Defoe's position. Acknowledging that 'the great man, to whom experimental philosophy next owed its obligations, was ... Francis Bacon, lord Verulam', Goldsmith writes:

> The ancients seem to have been but little acquainted with the arts of making experiments for the investigation of natural knowledge. It is true, they treasured up numberless observations, which Nature offered to their view, or which chance might have given them an opportunity of seeing; but they seldom went further than barely the natural history of every object: they seldom laboured, by variously combining natural bodies ... to *create* new appearances, in order to afford matter speculation.
> They were but little employed in thus diving into the secret recesses of Nature: they read the book as it lay before them ...
> (from 'Introduction' to *A Survey of Experimental Philosophy* (London, 1776) in Goldsmith 1966, V: 343–4)

> had no ships), even their physitians had no books. Experiments were the hight of their knowledge, and so we may suppose when a practiser had killed 4 or 500 he might pass for a doctor.
>
> (*Gentleman*, p. 67)

When Defoe states that no persuasion is either acceptable or necessary other than the plain arguments of evidence and demonstration, his position is that of the Baconian scientist. Locke had pronounced that 'Knowledge is seeing, and if it be so, it is madness to persuade ourselves that we do so by another man's eyes. ... Till we ourselves see it with our own eyes and perceive it by our own understandings ... let us believe any learned author as much as we will' (Locke 1922: 227). As a 'modern', Defoe declares, significantly, in the penultimate paragraph of his history of science: 'Let the Experiments be made and the Negative prov'd, and then indeed no Man will oppose it, for Demonstration puts an end to all Arguments; but till then we must be allow'd to judge as Reason and the nature of Things direct us' (p. 307).[23]

The *History of Arts and Sciences* has remained unexplored by the critics of Defoe, but there can be no question of its importance in establishing his concept of nature and of man; no adequate discussion of Defoe's intellectual indebtedness can afford to neglect it. Defoe demonstrates in this work his conscious alignment with the principles of New Science. The most decisive proof of his knowledge and esteem of the experimentalists' new approach occurs in his discussion of the magnet and loadstone. Here he observes that the discovery of the magnet can be put to 'innumerable' uses, and that some of these 'may be sum'd up out of the Learned Mr. *Boyl*'. The work that Defoe copies from is *Certain Physiological Essays* (London, 1661). Boyle had written:

> If on either of the extremes or poles of a good armed load-stone, you leisurely enough, or divers times, draw the back of a knife ... you may observe, that if the point of the blade have in this affriction been drawn from the middle of the aequator of the load-stone towards the pole of it, it will attract one of the extremes of an equilibrated magnetick needle; but if you take another knife ... thrust the back of the knife from the pole towards the aequator, or middle of the load-

---

[23] Defoe does not disregard the achievement of the ancients and in *An Essay upon Projects* he particularly praises 'the Romans [who] made it one of their principal cares to make and repair the highways of the kingdom' (p. 59).

stone ... it will not attract, but rather seem to repel or drive away that end of the magnetick needle, which was drawn by the point of the other knife.

(Boyle 1744, I: 218)

Here to compare is Defoe's text:

Mr. *Boyle* found by drawing the back of a Knife, or long piece of Steel Wire, &c. over the *Poles* of a *Loadstone* leisurely, once or divers time, beginning the Motion from the middle of the Equator of the Stone, towards the *Pole*, the Knife or Wire will accordingly attract one end of a poised magnetical *Needle*; but if you take another Knife or Wire, and thrust it leisurely over the *Pole*, from the *Pole* towards the Equator or middle of the Equator, this Knife shall drive or expel away the same end of the *Needle* which the former Knife would attract.

(p. 257)

Boyle concluded his experiment with the remark:

And this improbable experiment not only we have made trial of, by passing slender irons upon the extremeties of armed load-stones ... But whether, or how far this observation insinuates the operation of the load-stone to be chiefly performed by streams of small particles, which perpetually issuing out of one of its poles do wheel about and re-enter at the other; we shall not now examine.

(I: 219)

This passage is rendered by Defoe as

[this] Experiment makes it very probable that the Operation of the *Magnet* depends upon the flux of some fine Particles, which go out at one *Pole*, then round about, and in again at the other.

(p. 257)

Not only does Defoe 'sum up' Boyle's research but he imitates the style for recording experiments; even the way the scientist numbers and lists his experiments is repeated in the *History of Arts and Sciences*.[24]

The main function of Defoe's history of science is to give a comprehensive, accurate and up-to-date survey of the development of husbandry, navigation, trade and commerce 'in all Parts of the known World'. Defoe's ambitious aim is to trace chronologically the develop-

---

[24] And compare Boyle 1744, I: 284 with Defoe's experiment no. 23, p. 255. For other references by Defoe to 'the great and truly honorable Mr. Boyle' see *Gentleman*, p. 69, *The Storm*, p. 260, and *The Consolidator*, p. 280.

ment of 'all Branches of Knowledge' and so to collect intelligence of what has been done so far, as well as to provide guidance on what yet needs doing. The full title of the work reads:

> A General History of Discoveries and Improvements, in useful Arts, Particularly in the great Branches of Commerce, Navigation, and Plantation, in all Parts of the known World. A Work which may entertain the Curious with the view of their present State; prompt the Indolent to retrieve those Inventions that are neglected, and animate the diligent to advance and perfect what may be thought wanting.

We are reminded of Baconian histories 'for the Advancement of Universall Learning, and all manner of Arts and Ingenuities'. These encyclopaedic schemes, as William Petty judged, offered an opportunity 'whereby the wants and desires of all may bee made knowne unto all, where men may know what is already done in the businesse of Learning. What is at present in doing, and what is intended to be done' (Petty's evaluation of Hartlib's *Office of Address: Advice*, pp. 1–2).[25] While Defoe's scheme does not have the detail or scope of, say, Hartlib's registers, it does share the fundamental Baconian purpose of pooling and organising information to stimulate further research and improvements. Its goal was 'to animate', as Defoe has phrased it, 'the diligent to advance and perfect what may be thought wanting'. Behind Defoe's *History of Arts and Sciences* stands his conviction that the 'Encrease of Knowledge ... encreases the Felicity of Man's Life' (p. 2). It is Defoe's unshaken belief in the great material benefits derived from collaboration and communication of knowledge that makes him pronounce:

---

[25] And cf. 'An unpublished letter of Dr. Seth Ward relating to the early meetings of the Oxford Philosophical Society' where the Baconian stock-taking is explained thus:
> Our first businesse is to gather together such things as are already discovered and to make a booke with a generall index of them, then to have a collection of those wch are still inquirenda and according to our opportunityes to make inquisitive experiments, the end is that out of a sufficient number of such experiments, the way of nature in working may be discovered, but because ... we may probably spend our labour upon that wch is already done, we have conceived it requisite to examine all the bookes of our public library ... and to make a catalogue or index of the matters ... [so that] a man may at once see where he may find whatever is there concerneing the argument he is upon'.
>
> (Robinson 1949: 69)

> To *such*, [dispos'd to search for Improvements], this Work will, perhaps, be as Profitable, as Pleasant, at least, we shall give them in *Miniature*, and in Speculation, such practicable Things as they may not yet have consider'd, and as may some time or other rouse up Adventurers to undertake; that according to the undoubted design of that Providence which made the World, it may first or last be fully Improv'd, its Treasures fully Discover'd, and all that intrinsick Wealth which Heaven furnish'd the Globe with, be found out and made use of, as he certainly at first intended it shou'd be.
>
> (pp. 5–6)

Fulfilling the first principle of Bacon's philosophy, Defoe's history of science is directed to the benefit and use of man; in the words of the Preface, it is written '*for the benefit of the Ages to come*'. Defoe shares the Baconians' enthusiastic zeal for the advancement of 'fruitful' learning. He declares:

> *The same Zeal for the general improvement of the World which inspired those patrons of Wisdom to undertake great Things, prompts us to give our Age the History of their Discoveries, and Improvements; that Men of the same Genius may be encouraged from the success of former Times to pursue the like useful Discoveries for the benefit of the Ages to come.*
>
> (Preface)

By his own statement, the *History of Arts and Sciences* was prompted by an earnest desire '*to propagate useful Knowledge, for the good of Mankind*'. The place where Defoe first had encountered a similar compendious stock-taking of all arts and sciences was at the Dissenters' Academy. Morton's science lectures, the *Compendium Physicae*, were 'systems of all severall Arts and Sciences' (see pp. 38–42 above). While Defoe shifts the emphasis from a predominantly scientific to a social and commercial concern, he retains the quintessential Baconian goal of giving men

> from time to time, an inventory of what hath been already discovered; whereby the needless labour of seeking after known things may be prevented, and the progress of mankind, as to knowledge, might the better appear.

The words are Boyle's and come from *Certain Physiological Essays* (Preface), that is, from one of the works that Defoe definitely used and copied from when compiling his comprehensive *History of Arts and Sciences*.

The *Essay upon Projects*, *The Storm*, *The Consolidator*, *History of Arts and Sciences* and *The Compleat English Gentleman* were written over a period of more than thirty years – it is plain that Defoe's adherence to the ideals of New Science was not a passing whim. In these works we recognise the author's closeness to a wide range of Bacon's thought. Although Defoe may sometimes hide his intellectual indebtedness and scholarly sources, this does not prevent him from proclaiming (as already mentioned): 'the Fountain of all human Knowledge, and the great Mistery for the Wisest Men to search into ... [is] *Nature*', or 'upon the whole, the study of science is the original of learning; the word imports it. 'Tis the search after knowledge.'[26] In many of his fundamental beliefs Defoe is a Baconian.

[26] *History of Arts and Sciences*, pp. 238–9, and *Gentleman*, p. 211.

5

Defoe's *General History of Trade*: its relation to the Baconian histories

The idea of recording all branches of knowledge as they relate to husbandry, trade and navigation had been in Defoe's mind long before 1725, when he published his *History of Arts and Sciences*. In 1713 he began a work which promised to give 'a Succinct Account of all the Materials upon which the Arts and Inventions of Men have been employ'd, and from whence they have produced Manufactures'; Defoe entitled the project *A General History of Trade*.[1] In case the word 'history' should be misleading, he explains in the Introduction that he will not give a chronological account, instead his

> View is of another kind, (*viz.*) to shew you from whence our Trade is deriv'd, how it came to its present Magnitude, and what that present Magnitude really is; and for this reason I call it a *History*.
> (I: 20)

Once completed, Defoe writes, this work will be 'like a Map or Scheme in Miniature, of the whole World of Trade' (II: 4). In other words, Defoe uses the word 'history' here in the sense in which Bacon had used it for his Histories of Nature and of Trades. Like the Baconian histories, Defoe's *History of Trade* sets out to be a comprehensive, systematic study of relationships, providing guidelines for future inves-

---

[1] The *History of Trade* was issued in four instalments for the months of June, July, August and September. I shall refer to these by roman numerals, e.g. June: I, July: II etc., and quote them in the text.

tigations and improvements (see p. 11 above). In order to examine the full range and depth of Defoe's indebtedness, it will be necessary to give a brief outline of the seventeenth-century Histories of Trade.

The Baconian History of Trades

Bacon had considered the making of a History of Trades as by far the most important aspect of his programme for the regeneration of natural history. The use of the History of Trades, he had written, 'is of all others the most radical and fundamental towards natural philosophy', since it will 'be operative to the endowment and benefit of man's life' (Bacon 1857, III: 332–3).[2] It is worth noting that by 'trades' Bacon meant not trade in general so much as specific learned or technical skills. It was only later, in the second half of the century, that the word was used to refer to both, the work of the artificer and commercial exchange.

As I suggested earlier, the work in which Bacon described in greatest detail his idea of making the History of Trades is the *Parasceve, or Preparative towards a Natural and Experimental History*, affixed to the *Novum Organum* (1620). Stressing again the significance of the undertaking, he explained that 'the history of Arts is of most use, because it exhibits things in motion, and leads more directly to practice'. Of the histories to be compiled, Bacon advised that those are of greatest importance which 'exhibit, alter, and prepare natural bodies and materials of things; such as agriculture, cookery, chemistry, dyeing; the manufacture of glass, enamel, sugar, gunpowder, artificial fires, paper, and the like'. Next came 'those which consist principally in the subtle motion of the hands or instruments ... such as weaving, carpentry, architecture, manufacture of mills, clocks, and the like' (IV: 257–8). Bacon followed his instruction with a 'Catalogue of Particular Histories by Titles'. Of the 130 suggested titles or areas of future research no less than 48 (running from numbers 81 to 128) deal with the making of histories of 'trades'. So, for example, Bacon directed that there be a 'History of Baking, and the Making of Bread', 'of the Dairy', 'of the manufactures of Silk', 'of Flax, Hemp, Cotton, Hair', 'of Leather-making', 'of

---

[2] The following discussion of the Baconian History of Trades owes much to W. E. Houghton 1941: 33–60, C. Webster 1975: 424–7, and M. Hunter 1981: 91–112.

Pottery', 'of Basket-making', 'of Agriculture, Pasturage, Culture of Woods, etc.' (IV: 269–70). From the outset, then, Bacon urged that the method of the natural history programme be applied to technological and economic areas of society. Believing that it was in man's power to alter reality, he emphasised that the history of nature 'Wrought or Mechanical' offered the best opportunity for effecting this change.

Inspired by Bacon, men like Hartlib, Petty, Boyle and Evelyn set out to compile Histories of Trades. Indicative of their confidence in these schemes is Petty's claim (quoted in a letter from Hartlib to Boyle, dated 10 August 1658) that the History of Trades was one of 'the great pillars of reformation of the world' (Boyle 1744, V: 281). Combining two of Hartlib's favourite projects, the reform of society and the reform of education, Petty wrote in 1648 his *Advice of W.P. to Mr. Samuel Hartlib for the Advancement of some particular Parts of Learning*.[3] At the centre of this tract is the suggestion for the erection of 'a Gymnasium Mechanicum or a Colledge of Trades-men' for the promotion of 'all mechanical arts and manufactures'. This institution, Petty wrote, would offer 'the best and most effectuall opportunities and meanes, for writing a History of Trades in perfection and exactnesse' (p. 7). After expounding the 'Nature, Manner, and Meanes of Writing the History of Trades' he listed the 'Profits and Commodities' that we might expect from such an undertaking. Not only would it give an up-to-date inventory of all discoveries and inventions but 'a vast increase of honourable, profitable, and pleasant Inventions must needs spring from the work'. Such a scheme would eliminate misunderstandings and deceit, since 'all men whatsoever may hereby so look into all Professions, as not to be too grossly cozened and abused in them'. Furthermore, it would teach 'the study of Things' rather than 'the rabble of words'. Petty pointed out that men 'having so large a Booke of Gods works added to that of his word, may the more cleerely from them both, deduce the wisedome, power and goodnesse of the Almighty'. Reminding the reader that the 'History of Trades is also an History of Nature, but of Nature vexed and disturbed', Petty rounded off his essay with a reference to Bacon's 'Exact and Judicious Catalogue' given at the end of the *Parasceve*

---

[3] For Hartlib's continued interest in the History of Trades see p. 20 above.

(pp. 21–6). As Petty's 'Profits' clearly indicate, the History of Trades was closely bound up with Bacon's whole cast of thought.

Fundamental to the Histories of Trades is the belief that nature gives the materials, but art – human manufacture, inventions, discoveries – makes the products on which human life depends. This point was well made by Hooke.[4] Taking Bacon's 'Catalogue of Particular Histories by Titles' as his example, Hooke divided physical reality into 'The Histories of Natural things' and 'The Histories of Artificial and Mechanical Operations'. The second set of these titles is introduced with the observation that 'it will be requisite to take notice of, and enumerate all the Trades, Arts, Manufactures, and Operations about which Men are imployed'. Hooke proposed more than three hundred Histories of Trades, concluding that here 'inquisitive Persons that are ignorant of [trades], may come to a more perfect Knowledge of them'. More important still is the purpose of 'finding out the ways and means Nature uses, and the Laws by which she is restrain'd in producing divers Effects'. An encyclopaedic improving scheme, the History of Trades was drawn up for the intellectual as well as for the workman. Its aim was to give insight into the secrets of nature and to show the ways in which 'she may be assisted, accelerated, regarded, stopt, and the like' (Hooke 1705: 22–7). Returning to the subject in one of his 'Lectures concerning Navigation and Astronomy' (given 8 May 1685), Hooke pointed to the *raison d'être* of every Baconian historian. 'Art', he wrote,

> which at best does but mimick Nature, must search for materials where it can find them, and make use of such as can be procur'd ... So that tho' Art be far short of Nature in perfection of acting, yet since the power of it is placed in Man, it seems to be of as great a concern to him to be knowing and potent therein; for every new discovery therein gives him a new Power which he had not before.
> (Hooke 1705: 532)

No greater justification was needed for striving after the dominion over things than that God himself had placed in man the power 'to be knowing and potent' in the ways of nature.

---

[4] On 6 October 1664 Hooke wrote to Boyle: 'I am now engaged in a very great design, which I fear I shall find a very hard, difficult, and tedious task, and that is, the compiling a history of trades and manufactures ... by God's assistance, I shall endeavour to the utmost of my power, to go as far in it as I am able, being resolved wholly to apply my mind and endeavours to it' (Boyle 1744, V: 537).

By the middle of the century interest in these projects had become more widespread. From the records of the early meetings of the Royal Society it is evident that the History of Trades was a subject that frequently came up for discussion. On 16 January 1661 it was minuted that 'the catalogue of trades brought in by Mr. Evelyn, and that of Dr. Petty, were referred to them and Dr. Merret, to be compared, methodised, and returned to the society ... Mr. Evelyn was desired to bring in an history of engraving and etching: And Dr. Petty to communicate the history of some trade at his own choice.'[5] Three years later a committee for 'Histories of Trade' was inaugurated with Boyle, Evelyn, Hooke, Merrett, Petty, and Wilkins as members. Other related specialised committees that were founded in 1664 were the 'Mechanical' and 'Georgical Committee' (Birch 1756, I: 407). The work to be mentioned in connection with the last group is Evelyn's *Sylva, or a Discourse of Forest-Trees, and the Propagation of Timber in His Majesties Dominions* (London, 1664). Evelyn's work derived its inspiration from Bacon's suggestion for a 'History of Agriculture, Pasturage, Culture of Woods, etc.'; it was also influenced by the work carried out by the members of Harlib's circle. *Sylva* in many ways epitomises the Baconian scientists' efforts to link scientific discoveries with technology; not only did it encourage the improvement of agriculture and forestry, but, on the intellectual level, it stimulated further scientific research.

By 1667 Sprat recorded that the Fellows

> have propounded the composing a *Catalogue of all Trades, Works*, and *Manufactures*, wherein men are emploi'd, in order to the collecting each of their Histories: by taking notice of all the Physical Receipts, or Secrets, the Instruments, Tools, and Engines, the Manual operations or sleights, the cheats, and ill practices, the goodness, baseness, and different value of Materials, and whatever else belongs to the operations of all *Trades*.
>
> (Sprat 1959: 190)

Of the histories in progress at the time, Sprat mentions among others the 'History of the comparing of several *Soyls*, and *Clays*, for the better making of *Bricks*, and *Tiles*'; 'the examining' of the nature of *Petrifying*

---

[5] Birch 1756, I: 12. On 5 February 1661 Petty wrote to his brother that he had talked to the king for 'half an hour before the forty Lords, upon the philosophy of shipping, loadstones, guns, etc., feathering of arrows, vegetation of plants, the history of trades, etc., about all of which I discoursed *intrépide* and I hope not contemptibly'; see Fitzmaurice 1895: 104.

Springs'; 'the propagation of *Potatoes*'; 'the gradual observation of the growth of *Plants*, from the first spot of life; the increasing of *Timber*, and the planting of Fruit Trees'; 'the making Experiments with *Tobacco oyl*'; 'the probability of making *Wine* out of *Sugar-canes*'.[6] We see how well the History of Trades fulfilled the experimental scientists' aim of applying scientific method to meeting the everyday needs of life. If the public criticised the Fellows for being involved in trivial activities, they themselves were convinced that their histories of 'Nature and of Art' were directly applicable to the practical good of mankind (cf. M. Hunter 1981: 90).

By far the most detailed description of the History of Trades as an intellectual concept occurs in Boyle's defence of experimental science, *Some Considerations Touching the Usefulness of Experimental Natural Philosophy* (Oxford, 1663–71). Committed to the subject since his meetings with Hartlib in the 1640s, Boyle finally collected his ideas in an essay entitled 'That the Goods of Mankind may be much increased by the Naturalist's Insight into Trades' (1671).[7] Echoing Bacon, Boyle argued that scientific progress could only be attained if the philosopher and craftsman co-operate. Boyle's aim was 'to carry philosophical materials from the shops to the schools, and divulge the experiments of artificers, both to the improvement of trades themselves, and to the great inriching of the history of arts and nature' (III: 139). Rejecting the fastidiousness of the schoolmen, Boyle pointed out that often craftsmen 'unacquainted with books, and with the theories and opinions of the schools' will by their own 'sagacity or casual experiments' discover new and better 'uses and applications of things' (p. 168). Boyle referred to his own history of the trade of varnish and urged all the virtuosi of our country 'not to disdain to contribute their observations to the history of trades'. A good History of Trades, Boyle declared, was 'one of the best means to give experimental learning both growth and fertility' (p. 172).

This brief review of the Baconian Histories of Trades can only inadequately convey the vigorous interest and the faith of the virtuosi in these schemes. As I have tried to show, the method and aspirations of the Histories of Trades were intimately bound up with Bacon's whole system of thought. More than any other aspect of Bacon's

---

[6] Sprat 1959: 191–2 and 258–9 and see p. 110 below.
[7] *Usefulness ... The Second Tome*, III: 167–75.

programme of reform, these histories of 'nature at second hand' inspired belief in 'innumerable benefits to all practical *Arts*'. Sprat, acting as the mouthpiece for the Society's opinion, gives us an idea of the confidence the Fellows placed in the project in the mid-1660s:

> They have assured grounds of *confidence*, that when this *attempt* shall be compleated, it will be found to bring innumerable benefits to all practical *Arts*: When all the secrets of *Manufactures* shall be so discover'd, their *Materials* describ'd, their *Instruments* figur'd, their *Products* represented: It will soon be determin'd, how far they themselves may be promoted, and what new consequences may thence be deduc'd ... In short, by this help the worst *Artificers* will be well instructed, by considering the Methods, and *Tools* of the best: And the greatest Inventors will be exceedingly inlighten'd; because they will have in their view the labours of many men, many places, and many times, wherewith to compare their own. This is the surest, and most effectual means, to inlarge the *Invention*.
>
> (pp. 310–11)

Yet, despite this promising start, interest in the History of Trades began to diminish after 1670. A number of reasons account for this decline. The most important is this: the scope was simply too vast, even if it was a collaborative enterprise. In the third part of the century, science became more and more an elitist activity, and the Fellows' shift of interest from practical studies to scientific investigations for their own sake thwarted the development of the History of Trades. Although the Baconians of the mid-century frequently wrote as if there were a close collaboration between the intellectual and the artificer, this was in fact no more than an aspiration, an ideal that was very slow to be realised. The truth is that neither the scientist nor the workman easily parted with secrets or learnt from the other's profession (M. Hunter 1981: 92–112).

A notable exception to this general trend was John Houghton. A devoted Fellow of the Society, Houghton continued to be interested in '*useful Practice*'. In 1681 he published a periodical, *A Collection of Letters for the Improvement of Husbandry and Trade*, which included Evelyn's 'Panificium, or the Several Manners of Making Bread in France' (Evelyn's response to Bacon's advice for a 'History of Baking, and the Making of Bread'), as well as histories of the trades of malting and cider-making, on 'Improving Land by Marle', and 'on the great Improvement of

Land by Parsley'. Whereas Evelyn objected to 'conversing with mechanical capricious persons', Houghton published 'useful things fit for the *Understanding of a plain Man*' (J. Houghton 1692–1703, no. 1). Houghton's continued zeal for practical knowledge is also demonstrated in the transition he made from 'trades' to 'trade', remarking

> that what the Husbandman is concern'd for, is the *Materia prima* of all Trade; and that the finding a vent for his Commodities, is as necessary to his end, as it is to know the ways of Tilling, Planting, Sowing, Manuring, ordering, and improving of all sorts of Gardens, Orchards, Meadows, Pastures, Corn-Lands, Woods, and Coppices; as also of Fruits, Corn, Grain, Pulse, new Hays, Cattel, Fowl, Beasts, Bees, Silk-worms, etc. ... therefore I design not only to give Instructions for that end, but also the best accounts I can meet with, how they may be advantagiously parted with; which will necessitate me often to treat of such things as more strictly come under the second Head of my Title, *viz*. Trade.
>
> (J. Houghton 1681–3, I: 2–3)

At a time when scientists were no longer so keen on linking science and technology, Houghton continued to combine agricultural and technological investigations with contemporary economic problems. As such, he represents a vital link in the Baconian legacy as it was handed down to Defoe.

Defoe's *General History of Trade*

Outlining his undertaking, Defoe states that he will 'give a General view of the Situation of [the] Original Materials in the World, How they stand related to one another, and how all in especial manner to us in *Britain*' (II: 4). His intention is to investigate 'Trade, abstracted from all the separate Interest of Politicks, and the Concerns of Nation, as to Peace, War, Safety, Power, and the like' (II: 40) – an analytical interest that shows the work's scientific nature.[8] Quite a large section of each monthly instalment is taken up with commercial abstracts, ordered lists

---

[8] Defoe's original determination not to deviate into particularity could not be sustained. As he became more involved in the controversy over the commercial article of the Treaty of Utrecht, he became more engaged in particular topics of the day. It is with exasperation that he exclaims in the issue for September: 'It is Impossible to speak of Trade in General and leave out entirely the Trade to France, and I fear it will be impossible, at least to me ...' (IV: 3).

of natural products and manufactured goods. Dividing the products of nature into two categories, Defoe tables under the first heading those which 'are compleated by Nature for the immediate use of Mankind'. Here he lists 'Corn, Fruit, Plants, such as Drugs, Spices, Tea, Coffee, Sugar, with Metals, Minerals, Timber, Gold, Silver, Pearls, Jewels, Fish, Cattle, and the like' (III: 4). The second category comprises the 'first species' of nature which

> are in themselves useless to Mankind till they pass the various changes, necessary to transform them into other Shapes, and being only rough Materials are adapted to Use by the help of Labour, Industry, Art, and the Application of other Materials; the performing of which we call Manufacturing.
>
> (III: 5–7 and cf. I: 22–32)

Defoe deals here with nature improved 'by the Work of Mens Hands' – Bacon's 'nature wrought or altered'. He gives us a table of 'the first species' of nature perfected by 'the Operations of Mens Hands', an excerpt of which is reproduced here (see fig. 1).

As we have seen, the idea of looking back to the 'Materia prima' or 'first species' of things was very much part of the new philosophy. It demonstrated to the natural historian (and the historian) the close interdependence existing between nature and human art. Believing that art was an extension of nature, it was insisted that the history of 'trades' was also a history of nature, though of 'Nature vexed and disturbed' (Petty 1648: 26). The Baconians' guiding principle had been that while nature gives the materials, it is human manufacture that makes the products for the support of human life. Defoe's awareness of these ideas comes out strikingly when he writes that 'the Produce of Nature is indeed the *Materia Fabricata* of all Manufactures, but the Improvements of Art give a new Face to the very Species, so that you know it no more either by its Form, or by its Name' (*Brief State*, p. 10). In his *History of Trade* he points out that

> the Wooll, the Silk, the Flax, the Hair, all appear uncapable to serve to the Uses for which they had an Inherent fitness, till they had passed the various Operations of the Manufacturer; and were brought to such Degrees of Fineness, by Combing, Carding, Hackling, Dressing, Twisting, Throwing, Spinning, and so on.
>
> (I: 31)

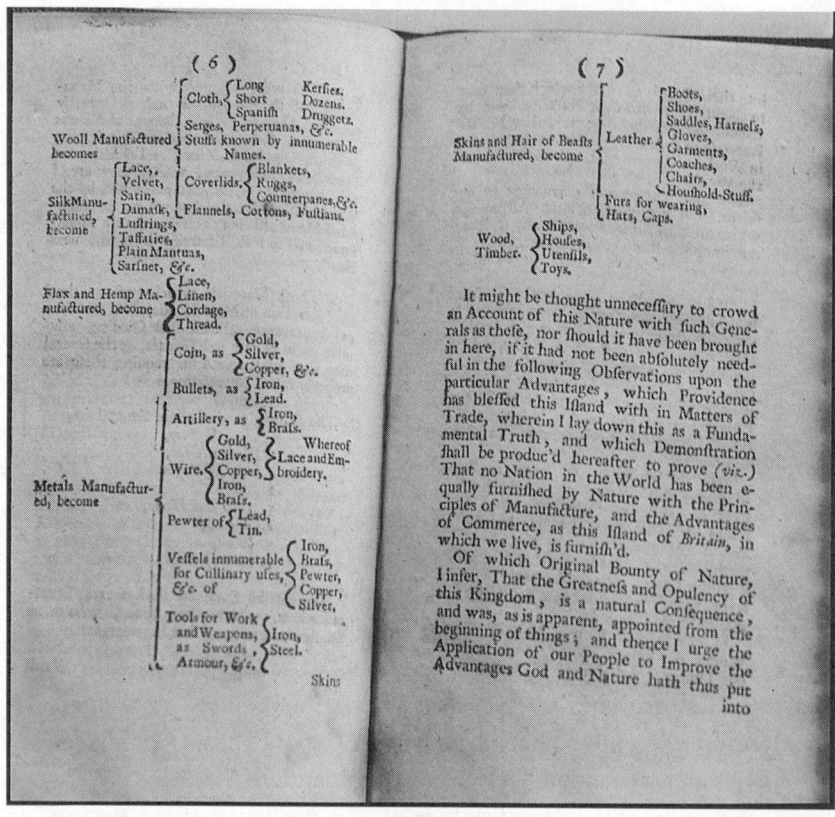

1   Defoe's 'Nature wrought and altered' from Defoe's *General History of Trade* (III: 6–7)

When Defoe points out that 'Corn was not to be Eaten whole' but altered and improved by human art 'for the support of Human Life', he echoes Hooke, who thirty years earlier had observed that art does 'help and promote Natures Operations; as it improves the Sight by Spectacles, Microscopes and Telescopes … It converts Corn into Bread' (Hooke 1705: 532). While man has not the power to create, he has the power to subject nature to his will. Like Bacon, Defoe observes that

> Nature having laid down these first Principles for the Wit and Invention of Men to work upon, Art has carried on so many Wonders of Improvement from them, that the Necessity of those first Principles seems to appear less to us than it did before.
>
> (I: 8)

Throughout, Defoe acknowledges a close connection between 'trades' and 'trade'. Broadly speaking, one can say that the first and third issues of the *History of Trade*, dealing with the origin of manufacture, stress the traditional Baconian interest in 'trades' or skills, while the second instalment focuses on trade and commerce. But the distinction is not a clear-cut one. With John Houghton, Defoe argues that investigations of the 'Materia prima', of manufactured goods and of trade, are interrelated; one cannot discuss one without reference to the other.

The second issue of the *History of Trade*, then, deals with 'the order of Place where those first Principles of Trade ... are Deposited' (II: 6). Defoe registers what every country of the world produces, confining himself, as he informs us, to what is particular to a country. He 'travels' from country to country assembling and ordering all that is produced for use, benefit and pleasure. In his intention to reflect the complete cycle of trade throughout the world, he carries out in the world of trade what the Baconians had aimed at in the world of arts and craft techniques, namely, 'to compile, and publish a compleat Cycle and History of Trades, with whatsoever else he should judge of use and benefit of mankind'. For his *History of Trade* Defoe relies not only on books and maps, he uses personal experience as well as his imagination. Ten years later, though, he published the results of actual tours through Great Britain, where he had been 'collecting with a most insuperable diligence all that the mechanics had invented for Agriculture, Architecture, and the fabric of all sorts of works, belonging to sports, and to cloathes, for use and for magnificence'. These words of, John Evelyn, describing the Baconians' contribution to a History of Trades, match Defoe's intention for the *History of Trade*. In method and aim this earlier and incomplete work prepares us for one of Defoe's masterpieces, his *Tour thro' the Whole Island of Great Britain* (1724–6).[9]

The aim of Defoe's classification of the world of trade is to reflect the 'proper Order' in which 'Nature has placed' the natural resources of the world (III: 3). Behind Defoe's manifest commitment to mapping the world of trade stands the Baconian conviction that there is an order in the creation which man can discern and reproduce in his histories of

---

[9] From John Evelyn's 'An Account of Signor Giacomo Favi'; see Evelyn 1825: 248 and 250.

nature and of man. If we study the earth, the multitude of 'Inlets, Gulphs, Bays, Firths, deep Channels of Rivers, and the like, they *seem* to present a kind of Confusion, distort the Figure of the Earth, and *look like* a Deformity of Parts' (II: 31, my italics). Defoe refers us to Thomas Burnet's *Sacred Theory of the Earth* (published in Latin in 1681 and 1684, in English in 1684 and 1690), where we read: 'if we consider the whole surface of [the earth], or the whole Exteriour Region, 'tis as a broken and confus'd heap of bodies, plac'd in no order to one another, nor with any correspondency or regularity of parts'. In Burnet's judgement the earth is 'a great Ruine, and [has] the true aspect of a World lying in its rubbish' (Burnet 1684: 110). That Defoe mentions Burnet's *Sacred Theory* only to refute it becomes clear in the following paragraph, where he declares with emphasis:

> Yet has all this Confusion and appearing Dislocation, the greatest Beauty and Harmony imaginable in it, if we consider how all these are directed before hand, by the secret Power who foreknew the Occasion; how without these, Navigation, which was to be the great medium of Commerce, in the World, would be impracticable; how, if the Land had been every where surrounded with steep Rocks, high Cliffs, and unapproachable Hills, or had run out with continued Flats and Sands far into the Sea, Ships might have been Built indeed, but they would have found it very difficult to have gotten them into the Sea.
>
> (II: 32)

Defoe's aim is to prove, in opposition to Burnet, that there is a wise order in this 'appearing disorder'; his *History of Trade* seeks to reflect it.

His argument is that not a revengeful but a loving God has dispersed all the parts of the world, so creating the need for trade and exchange. Nature 'in its first Directions was commanded to fit the World for Trade' (II: 30). 'Had every Nation enjoy'd in it self every thing needful for the Life, Pleasure, Wealth, and Strength of all its Inhabitants', no commercial intercourse would have been necessary. 'But Providence has cut out the World for other work, and the wise Creator has most evidently shewn to us, that he had design'd the World for Commerce, from the measures taken in forming the Globe, in appointing Seasons, varying the Productions according to the difference of the Climate, the Soil, and the position of the parts' (I: 9–10). With John Evelyn, devotee of the New Science and proposer of a History of Trades, Defoe believes that the earth 'from the very Beginning, [was] dispos'd for Trafick and Commerce'. Writing in 1674, Evelyn declared that the earth 'presents us with a thousand Objects of Utility and Delight'. And

> though, through her rugged and dissever'd Parts, Rocks, Seas and remoter Islands, she seem at first, to check our Addresses; Yet, when we ag'en behold in what ample Baies, Creeks, trending-Shores, inviting Harbours and Stations, she appears spreading her Arms upon the Bordures of the Ocean; whiles the Rivers, who re-pay their Tributes to it, glide not in direct, and precipitate Courses from their Conceil'd, and distant Heads, but in various flexures and Meanders ... methinks she seems, from the very Beginning, to have been dispos'd for Trafick and Commerce, and even Courts us to visit her most solitary Recesses.
>
> (Evelyn 1674: 2)

Defoe's certain acquaintance with these ideas is demonstrated in the passage quoted above. We see it again in his observation that every inquiring mind must see how 'wonderfully the Blessings of the Creation are Disperst up and down, and how duly proportion'd for the Benefit of the whole; How assistant to one another, how happily proportion'd to the Advantages and to the Disadvantages of the People who Inhabit the respective Parts' (I: 24).

Defoe does not find these religious reflections out of place in a work on trade. On the contrary, he considers the study of secular knowledge as 'the most profitable and useful way of conveying Sacred Knowledge, and Improving both our selves and others'. 'As to the Propriety of such Digressions', Defoe has this to say: 'It is very much my Opinion, that due Observations of the Wisdom, Foreknowledge, and Omnipotence of God, should run through all our Discourses of Civil Affairs, and be the Constant Application of every Branch, as well of our Writings as Conversation' (I: 24–5). Again and again he bursts into enthusiastic admiration of the Creator's divine plan which has directed 'the common Business of the World [to be] subservient to the glorious Design of God'.

> Thus the Riches of the World, the Valuable Drugs, the useful Dying Stuffs, and Dying Woods, the Sugars, Tobacco, rich Furs, and the Fish, as well as Gold and Silver, are by the wise appointment of the

> Great Disposer of all things, Dispersed over the whole Country, to make the Nations... oblig'd to visit them.
>
> (II: 17–18)

And again:

> Thus the Wise Disposer, has separated all those valuable things, by vast Oceans, unknown Gulphs, and almost impassable Seas, that he might joyn them all again, and make them common to one another, by the Industry of Men, and thereby propagate Navigation, Plantation, Correspondence, and Commerce to the Universal benefit of every part of the World.
>
> (II: 25–6)

This is by no means an exhaustive list of Defoe's statements praising the divine order in nature, but it is representative of his characteristic emphases. We recognise in him the Christian virtuoso who, studying nature, is struck with the perfection of God's work. With Locke, Defoe might have exclaimed that 'the works of Nature everywhere sufficiently evidence a Deity'.[10] While Ray, Hooke, Boyle and Wilkins argued that the wisdom of God was revealed by science, Defoe, adopting and adapting this belief, argues that Providence is revealed by trade. His *History of Trade* teaches to 'Honour the Wisdom of Providence' – one of the 'Profits' detailed in Petty's outline for a History of Trades was to 'deduce the wisedome, power and goodnesse of the Almighty'. 'And what if', Defoe asks in his *Review* essay written in February 1713 (a few months before the publication of the *History of Trade*), 'I should tell you there is a kind of Divinity in the Original of Trade, and if I spend one Review to put you in Mind that God, in the Order of Nature, not only made Trade necessary to the making the Life of Man Easy ... but also qualified, suited and adapted the Vegetative and Sensitive World to be subservient to the Uses, Methods and Necessities which we find them now put to, by the ingenious Artists, for the Convenience of Trade?' (IX: 107).

Trade, in Defoe's thinking, is a bond that knits mankind together. It is thanks to trade that 'every improving quality [is] circulated through the World, and the whole Globe seems now to be brought into a kind of general acquaintance with it self, the remotest Nations converse, the People know one another, nay, I may say, continually talk with one

---

[10] Willey 1980: 27.

another, by Missives, by Messengers, and by Correspondencies of all sorts' (*History of Arts and Sciences*, pp. 79–80). The chain of trade provides a unifying element in social life. It makes the country and the city dependent upon each other and the rich indebted to the poor, for 'were it not for the said Poor, the Rich would go Naked ... And on the other Hand, the Poor are so much beholding to the Rich, that were it not for these Labours ... they would be Starv'd' (I: 29). The efforts of the husbandman, tradesman and craftsman, Defoe insists, must be 'JOYN'D, I desire you would mark the word, for I cannot but observe here, That unless these had all JOYN'D' in the past, 'the present State of Wealth and Greatness' could never have been achieved (I: 14–15). On the same topic of collaboration, Defoe remarks that

> Trade could not have done its part without the Adventurous Sailor, the Industrious Artificer, the Expert Manufacturer, to Build, Furnish, Freight, and Navigate the Shipping we Trade with. The Lands could not have done their part, without the Laborious Husbandman to cultivate and improve them, and without the substantial Farmer and Grazier.
>
> (I: 15)

All members of society – but especially the practical 'experienc'd *Men*' – contribute to the harmony and wealth of the nation. Defoe argues that the secret of national wealth lies in the chain of diverse links joining the apparently insignificant with the larger trading world. In one of the last works to be published in his life-time, he reflects on the manufacture of a pin:

> Thro' how many Hands does every Species pass? What a Variety of Figures do they Form? In how many Shapes do they appear? From the Brass Cannon of 50 to 60 hundred Weight, to half an Inch of Brass Wire, called a *Pin*, all equally useful in their Place and Proportions?
>
> On the other Hand, how does even the least Pin contribute its nameless Proportion to the Maintenance, Profit, and Support of every Hand, and every Family concerned in those Operations, from the Copper Mine in *Africa*, to the Retailer's-Shop in the Country Village, however remote?
>
> (*Brief State*, p. 13 and cf. *A Brief Case of the Distillers and of the Distilling Trade in England* (1726), p. 7)

Some critics have seen in Defoe's idea of the 'harmony of trade' an adumbration of Adam Smith's theory. It seems to me more profitable to note his debt to the seventeenth century. Closely related to Bacon's whole cast of thought, Defoe's theory of the chain of trade reflects a concept of nature and of man that we especially encounter in the project for a collaborative History of Trades.[11]

Starting from Bacon, Boyle in his History of Trades had suggested that the philosopher benefits from an exchange with the 'uneducated' craftsman. Never before had the artificer been elevated to the status of the philosopher. In Defoe's *History of Trade*, with its decisive shift of attention, it is the merchant on whom this honour is conferred. Not science but commerce is celebrated as working towards the 'Encouragement of Art, Science, and Human Wisdom in the World' (I: 24). For Defoe it is not the curious scientist but the 'inquisitive Merchant [who] searching every corner of the World' brings back facts and experience to augment the knowledge of the nation. He elevates trade to the high place previously held by philosophy when he describes it as 'a Patron of Arts, as it is the Mother of Industry; Commerce is naturally an encourager of Learning, and has by its Correspondence been the greatest assistance to human Knowledge' (*History of Arts and Sciences*, p. 80). If man stands to recover the dominion over things, this will, in Defoe's view, be due to the knowledge and experience of the craftsman and merchant.

Although the centre of attention has changed, the fundamental commitment of the Baconian historian is retained. This historian strove to instruct so that 'inquisitive Persons that are ignorant ... may come to a more perfect Knowledge' (Hooke). Defoe declares that 'one of the Great Ends of this Undertaking' is 'to rectify all the Mistakes which we are now fallen into in Trade, and to Inform those who are willing to Enquire into the Truth and Nature of things' (I: 5). While his predecessors compiled their compendious schemes for the 'laying open of the Mystery of Trades', Defoe's aim is to give insight into the

---

[11] It is a view that we find echoing through Addison's *Spectator*. In no. 69, for 19 May 1711, Addison sees nature as having 'taken a peculiar care to disseminate the blessings among the different regions of the world, with an eye to this mutual intercourse and traffic among mankind ... there are not more useful members in a commonwealth than merchants. They knit mankind together in a mutual intercourse of good offices, distribute the gifts of nature, find work for the poor, and wealth to the rich, and magnificence to the great.' And see the helpful article by Sawday 1983.

'mystery of Trade' itself. His purpose is 'to move, perswade, and instruct' the British nation to 'Diligence and Application, to make use of the Advantages, which, it is manifest we enjoy' (III: 8). Not 'the melioration of the professions' but the 'Improvement of Trade' is his aim (I: 42–3, 45; III: 7, 9, 45). Convinced that a detailed understanding of trade will lead to future improvements, he wishes to make the British people 'knowing and potent' therein, for 'every new discovery therein gives [them] a new Power which [they] had not before' (from Hooke, see p. 84 above). While knowledge gives power, blind ignorance leads to selfishness and destruction. Ignorance of the interaction of the various branches of trade have caused that 'our Harbours lye neglected; our Rivers not made Navigable; our Lands Uncultivated; our Collonies Unimprov'd; our Manufactures not Regulated; our Trading Privileges Unequally Granted; our Corporations Ungovern'd; our Poor Unemploy'd; Clandestine Trade Unsuppress'd; the fair Merchant Unencouraged' (I: 13).[12] Defoe may paint an unnecessarily bleak picture of England's social and economic life, but the message is clear. Assessing the importance of his self-imposed task as teacher, he pronounces that

> no Man can do his Country a greater Service, than to open their Eyes, and encourage their Hands to Industry and Improvement; to let them see how they are Furnished, by God and Nature, with Materials for Commerce.
>
> (III: 43)

He confesses that he was challenged to this goal by 'a Regard to Publick Good'. For the present investigation of Defoe's alignment with the principles of New Science, his repeated statements of devotion to the advancement of 'useful Knowledge' are revealing. Comparing his *History of Trade* to a beacon of light, he hopes that it will guide and direct

---

[12] And cf. Hartlib's *Legacy* (London, 1651) where we read that '*Deficiency, is the Ignorance of the Husbandry of other places (viz).* what *seeds*, what *Fruits*, what *Grasses* they use; what *Ploughes*, *Harrowes*, *Gardening-tooles* they have ... dayly new Plants are discovered, useful for *Husbandry*, *Mechanicks*, and *Physick*, and therefore let no man be discouraged, from prosecuting new and laudable *Ingenuities*. And I desire *Ingenious Gentlemen* and *Merchants*, who travel beyond Sea, to take notice of the *Husbandry* of those parts ... And I intreate them earnestly, not to thinke these things too low for them, and out of their callings: nay I desire them to count nothing *triviall* in this kinde, which may be profitable to their *Countrey*, and advance knowledge. And truly, I should thanke any *Merchant* that could informe me in some *triviall* and ordinary things done beyond Sea' (pp. 78–81; and see pp. 169–70 and 173–4 below).

the British nation and will be 'a Work of publick Benefit, and Useful to all that shall think fit to give it the Reading' (II: 35–6). A decade or so later, he expresses the same commitment when he writes that he was '*prompted*' by an earnest desire '*to propagate useful Knowledge, for the good of Mankind*' (*History of Arts and Sciences*, Preface and see pp. 169–74 below).

So far the link between the seventeenth-century History of Trades and Defoe's *History of Trade* has been little remarked. Yet anyone familiar with the period will see in his schematisation of the world of trade a perfect reflection of the experimentalists' ideology. Defoe's original concept of his *History of Trade* resembles in ideas, structure and aim the material which entered into its seventeenth-century predecessors, and extends that tradition.

The New Sciences exerted a profound and pervasive influence upon Defoe. In works like the *History of Trade*, the *History of Arts and Sciences*, *A Plan of the English Commerce* (1728), *Humble Proposal to the People of England* (1729) or *Brief State* (1730) we find evidence of his intimate familiarity with the principles of the new philosophy. In these relatively little-read texts on trade we gain insight into his concept of the world of nature and of man. Under the influence of the New Sciences, Defoe advocates the value of personal observation and personal experience, and the ideal of sharing; he stresses the knowledge of *things* rather than of *words*. In these works he declares his enthusiastic desire to promote useful knowledge for the good of all. As a Baconian he sets out to convince his age that it is man's God-given duty to explore and use nature for the relief of man's estate. He diligently observes, collects, compiles all that agriculture, trade and industry have invented for the use of mankind, and he goes on to make suggestions for improvements 'for the benefit of the Ages to come'.

Defoe's non-fictional tracts, heavily Baconian in character, help us to understand his fictional works. For it is here, especially in the works on trade written just before, during and after Defoe's enormous output of fiction, that we are best prepared for Crusoe's, Moll's or Roxana's view of the world. Our awareness of Defoe's thorough understanding of the Baconian vision of man's empire over things gives us the point of view from which to observe Crusoe's dominion over nature, as the following chapter will show.

6

Robinson Crusoe: man's progressive dominion over nature

The world of nature and of man

Defoe's views on the New Sciences are expressed throughout his long career as a writer, and he undoubtedly shared their goals and methods. In *The Life and Strange Surprizing Adventures of Robinson Crusoe* (1719), I shall argue, the characteristic Baconian belief in man's duty to study, alter and improve nature to his various uses is rendered in fictional terms. It is especially during Crusoe's solitary stay on the island that these fundamental tenets of experimental science emerge most clearly. My central focus will therefore be on the island experience in the original volume of Crusoe.[1]

Cast up on his island, Crusoe finds himself 'reduced to a meer State of Nature'. He makes his way to the shipwreck and resolves to 'pull every Thing to Pieces that I could of the Ship, concluding, that every Thing I could get from her would be of some Use or other to me' (p. 84). He builds a raft and gradually brings to his island provisions, clothes, pens, ink and paper, a gun, ammunition, essential tools and

---

[1] The literature on *Robinson Crusoe* is, of course, immense; here I shall refer only to material that is specifically relevant to my topic. For the first volume I have used the edition of J. Donald Crowley, World's Classics paperback (Oxford,1981), for volume three, *Serious Reflections during the Life and Surprising Adventures of Robinson Crusoe* (1720), I followed Aitken 1895. Page references are supplied in the text.

even some scientific instruments, such as 'three or four Compasses, some Mathematical Instruments, Dials, Perspectives [that is, small telescopes], Charts, and Books of Navigation' (p. 64). 'Had the calm Weather held,' he tells us, 'I should have brought away the whole Ship Piece by Piece' (pp. 56–7). Crusoe is 'reduced to a meer State of Nature', but the ship contains the artefacts of human culture, which he dismantles, brings ashore and reassembles bit by bit. Surveying his effort, he considers that he has 'the biggest Maggazin of all Kinds now that ever were laid up ... for one Man' (p. 55). In other words, Crusoe has been granted a basic stock of the standard products of seventeenth-century science; he begins his island existence equipped with the prerequisites for the world of culture that he is to create. It is not in vain that he reflects: 'What would have been my Case, if I had been to have liv'd in the Condition in which I at first came on Shore, without Necessaries of Life, or Necessaries to supply and procure them?' (p. 63).

However, the materials alone would not have been enough to guarantee Crusoe's survival. Defoe endows his hero with a keen desire to explore, experiment and learn from experience. An abundance of evidence illustrates Crusoe's objective and mathematically precise approach to reality. His first entry in his diary reads: 'It was, by my Account, the 30th. of Sept. [1659] when, in the Manner as above said, I first set Foot upon this horrid Island, when the Sun being, to us, in its Autumnal Equinox, was almost just over my Head, for I reckon'd my self, by Observation, to be in the Latitude of 9 Degrees 22 Minutes North of the Line' (pp. 63–4). He records with scientific exactitude the place and method of the pitching of his tent:

> This Plain was not above an Hundred Yards broad, and about twice as long, and lay like a Green before my Door ... It was on the *N.N.W.* Side of the Hill, so that I was shelter'd from the Heat every Day, till it came to a *W.* and by *S.* Sun, or thereabouts, which in those Countries is near the Setting.
>
> Before I set up my Tent, I drew a half Circle before the hollow Place, which took in about Ten Yards in its Semi-diameter from the Rock, and Twenty Yards in its Diameter, from its Beginning and Ending.
>
> In this half Circle I pitch'd two Rows of strong Stakes ... the biggest End being out of the Ground about Five Foot and a Half,

and sharpen'd on the Top: The two Rows did not stand above Six Inches from one another.

(pp. 58–9)

'By Observation', as Crusoe has just informed us, he reckons the correct time and place of his arrival on the island; it is 'by Experience' and 'Observation' that he will later learn about the changes of the seasons and the weather, as well as 'the Ebbing and the Flowing of the Tide' (pp. 106 and 150–1). Being a careful observer and experimenter Crusoe succeeds in making pots, candles, tools, clothes, shoes – indeed everything needed for his daily existence. 'I try'd many Ways to make my self a Basket', he tells us, and recalls that in his youth he used to watch a basket-maker, and, being *'a great Observer of the Manner how they work'd those Things ... I had by this Means full Knowledge of the Methods of it*, that I wanted nothing but the Materials'. Crusoe's early aptitude to observe how things are made now stands him in good stead, and with the right twigs he soon 'employ'd [him] self in making ... a great many Baskets' (p. 107, my italics).[2]

There is a correct pattern of procedure, starting by being critically alert and judicious. In Crusoe's own words: 'by stating and squaring every thing by Reason, and by making the most rational Judgment of things, every Man may be in time Master of every mechanick Art' (p. 68). The negative example, employing 'a most preposterous [that is, back-to-front] Method' of procedure, is provided by his initial construction of a boat. Quite irrationally he decides to build a large boat that could have held twenty-six people – a job, needless to say, far too big for one pair of hands, and in any case pointless (why is it so big?). Here Crusoe's otherwise thoroughly rational view of the world breaks down, and 'like a Fool', he realises, he had 'never once consider'd how [he] should get it off of the Land' (p. 126). In the end he is forced to let the unlaunched boat rest where she lies, using her 'as a *Memorandum* to teach [him] to be wiser next Time', to 'count the Cost' and 'judge rightly' beforehand the practicality of his design (pp. 128, 136).[3]

---

[2] Young Defoe's keen interest in his surroundings is well known, see Sutherland 1950: 1–25, 242, Moore 1958: 20–7, and Bastian 1981: 18–31.

[3] See Bacon III: 223. Writing on the proper evaluation of labour and exertion, Bacon referred his readers to Solomon, who 'saith excellently, *The fool putteth to more strength, but the wise man considereth which way*'. Defoe on a number of occasions mentions Solomon's fool as, for example, in his *History of Arts and Sciences*, p. 232 and *Gentleman*, pp. 66 and 100. In *Crusoe* 3 he contrasts

Obviously, the significance of this account lies in Crusoe demonstrating to himself, and to us, the inestimable value of making rational observations. His habitual method of procedure (and this exception merely serves to prove the rule) is the 'trial and error' method of the experimenter: it assists him in 'mastering this Difficulty' as well as every other that crops up during his stay on the island.

Nowhere is Crusoe's belief in observing and experimenting more clearly demonstrated than in the episode of his accidental crop. Rummaging one day in his things, he finds a bag; he carelessly empties its contents near his fortification and then forgets about the incident. A month later, he discovers to his surprise a few stalks of 'something green', which he identifies as barley. When the corn is ripened, he carefully collects the ears and sows some of them – not one of the seeds begins to grow. This, Crusoe confesses, 'was one of the most discouraging Experiments that I made at all', for 'I bought all my Experience before I had it.' The ensuing observation on his failed crop indicates his habit of mind: 'I lost all that I sow'd the first Season, by *not observing* the proper Time; for I sow'd it just before the dry Season, so that it never came up at all' (pp. 104 and 79 respectively; my italics). Realising that he should have paid more attention to the weather, he accordingly begins to register and table weather changes. In due course he deduces from his 'Observations' that he can count on a rainy and a dry season, and that the two alternate with reliable regularity. With this newly gained knowledge, he decides 'to make another Trial'. He now sows his seeds in 'the rainy Months of *March* and *April*', and finds to his great satisfaction that it 'sprung up very pleasantly, and yielded a very good Crop'. Painstaking experiments are at last rewarded with success as Crusoe reports that

> by this Experiment I was made Master of my Business, and knew exactly when the proper Season was to sow; and that I might expect two Seed Times, and two Harvests every Year.
>
> (pp. 104–5)

Crusoe defends an attitude to reality that can be directly related to the Royal Society.[4] Taking up the traditional Baconian interest in mechan-

---

his hero's conscientious search for knowledge with 'Solomon's fool [who] hates knowledge' (pp. 176 and 187). For a further discussion see p. 116 below.

[4] Also worth noting in this context is the fact that Crusoe chronicles both his successful and his

ical arts, the Fellows had stressed that scientific discoveries and inventions must be directed to everyday needs of life. Boyle, an enthusiastic defender of the Baconian History of Trades, had declared that he would not dare to think himself 'a true naturalist, till my skill can make my garden yield better herbs and flowers, or my orchard better fruit, or my field better corn, or my dairy better cheese, than theirs that are strangers' to experimental science (Boyle 1744, I: 463).[5] Hooke, another devotee of the History of Trades, had advised that 'We should therefore endeavour to be acquainted with all the various mechanical ways of Hammering, Pressing, Pounding, Grinding, Rowling, Cutting, Sawing, Filing, Steeping, Soaking, Dissolving, Heating, Burning, Freezing, Melting, and the like.' Another help, Hooke advised, would be to study *how many various mechanical ways there may be of separating Bodies joyn'd and mixt.* Such as Winnowing, Straining, Wringing, and Pressing, Washing, Distilling, Evaporating, Precipitating, Subliming, Crystallizing, Burning, Coppelling [that is, joining metals], Freezing, Shaking, Knocking, and the like.'[6] The Baconian scientists' belief in man's need to achieve dominion over things is exemplified in Boyle's pronouncement that 'man's power over the creatures consists in his knowledge of them; whatever does increase his knowledge, does proportionately increase his power' (III: 155 and cf. Bacon 1857, IV: 32).

One of the main concerns of chapter 5 was to show that Defoe shared this conviction. In his *History of Trade*, as in *An Historical Account of Sir Walter Raleigh*, the *Tour*, *History of Arts and Sciences*, *Compleat English Gentleman*, *Plan of Commerce*, *Humble Proposal to the People of England*, and *Brief State of the Inland and Home Trade* Defoe declares his belief in man's power to harness the forces of nature and make them subservient to his needs (see pp. 88–91 and 97 above). In *Crusoe* Defoe translates this

---

failed experiments, and is shown learning from his mistakes. Because of his complete report, we can testify that his knowledge is derived from and validated by what the experimental scientists called 'the diligent pursuit of truth'. Cf. Boyle's '*Two Essays* concerning the Unsuccessfulness of Experiments', Boyle 1744, I: 204–27.

[5] Referring to the Society's already begun histories, Sprat recorded among others those for 'the propagation of *Potatoes*'; 'the gradual observation of the growth of *Plants*, from the first spot of life'; 'the increasing of *Timber*, and the planting of Fruit Trees' and many others, Sprat 1959: 191; see pp. 85–6 above.

[6] This is only a small excerpt from the long list of craft-techniques given by Hooke: see Hooke 1705: 24–6 and 59–61.

concept of nature and of man into the 'real' world of fiction. As we watch Crusoe plant, harvest, grind corn, bake bread, make pots and baskets, build his fortification, and so on, we watch him command nature, bringing her 'to be serviceable to [his] particular Ends' (Boyle).

Crusoe's account is a demonstration of the experimental scientists' preoccupation not just with making things, but with human skills promoting the works of nature. His harvest brought in, he reflects on all the interdependent stages: 'the strange multitude of little Things necessary in the Providing, Producing, Curing, Dressing, Making and Finishing this one Article of Bread'. When the corn is ripened, he lists 'how many things [he] wanted, to Fence it, Secure it, Mow or Reap it, Cure and Carry it Home, Thrash, Part it from the Chaff, and Save it ... a Mill to Grind it, Sieves to Dress it, Yeast and Salt to make it into Bread, and an Oven to bake it' (p. 118). Reading Crusoe's lists of interrelated 'trades' is like reading an index to the Baconian Histories of Trades.

Most challenging of all was the making of earthen pots. Crusoe describes the stages of production, and tells us 'how after having labour'd hard to find the Clay, to dig it, to temper it, to bring it home and work it', he could only make some comical 'ugly things' (p. 120). He improvises a kiln to fire these and succeeds in making some tolerable pots: 'After this Experiment, I need not say that I wanted [that is, lacked] no sort of Earthen Ware for my Use.' His next concern is to get 'a Stone Mortar, to stamp or beat some Corn in'. To supply this want he is at first at a great loss, since 'for all Trades in the World I was perfectly unqualify'd for a Stone-cutter, as for any whatever'. Eventually having found a block of hard wood, and '*made* a hollow Place in it', he then '*made* a great heavy Pestle or Beater, of the Wood call'd the Iron-wood'. His next problem is how 'to *make* a Sieve, or Search [that is, a searce or strainer] ... to part [the corn] from the Bran, and the Husk', and he finds a solution for this, too. Although, 'to be sure I had nothing like the necessary Thing to *make* it', he remembers some muslin scarfs rescued from the shipwreck and with these he '*made* three small Sieves'. When it comes to the 'baking Part' and how he 'should *make* Bread', Crusoe finds he has no yeast, but as there is no way of improvising this need he wastes no time on trying. 'But for an Oven, I was indeed in great Pain; at length I found out an

Experiment for that also.' It is in his fourth year on the island that he baked his first 'Barley Loaves, and became in little Time a meer Pastry-Cook into the Bargain; for I *made* my self several Cakes of the Rice, and Puddings' (pp. 121–3; I have italicised the repeated uses of the words for making).

Crusoe is *homo faber*, the maker of things. Similar episodes using clusters of the verb 'to make' and its derivatives are found in connection with the construction of tools, weaving of baskets, in an earlier account of the baking of bread, and in the description of Crusoe's tailoring of clothes (pp. 73–4, 107, 117–18, 134–5; for a further discussion see p. 127 below). No sooner has Crusoe solved a problem or reached a peak of knowledge than a whole series of new problems come into sight, which, with indefatigable diligence, he proceeds to solve. 'Every new discovery ... gives him a new Power which he had not before' (Hooke 1705: 532). *Crusoe* may stand as an allegory of the advancement of learning: a model of initiative and invention. As a true Baconian, Crusoe 'endeavours to be acquainted with all the various mechanical ways of Hammering, Pressing, Pounding ... and the like'. The Baconians had argued that 'Nature puts us into the World more naked than most other of our fellow Creatures; but Art has abundantly supply'd that Defect' (Hooke 1705: 532). Crusoe's story of finding himself (almost) naked on an island, and supplying his defects with his efforts to improve himself in every mechanical art, takes on a new meaning when read in connection with the Baconian natural histories of 'trades'.

Over the years Crusoe becomes a baker, cheese-monger, saddler, stone-cutter, potter, carpenter, shipwright, tailor, basket-weaver, farmer and hunter. In Crusoe's inquiries into 'the mysteries and practices of these trades', we have Defoe's response to the Baconians' demand for histories 'of Baking, and the Making of Bread', 'of the Dairy', 'of Leather-making', 'of Stone-cutting', 'of Pottery', 'of working in Wood', 'of Basket-making', 'of Agriculture, Pasturage, Culture of Woods', 'of Gardening', 'of Hunting and Fowling' etc. (from Bacon's advice for Histories of Trades, IV: 269–70, and see pp. 81–4 above). When Defoe had first explored these ideas six years earlier, in his *History of Trade*, he had hoped to compile 'a Map or Scheme in Miniature, of the whole World of Trade'. This plan was never completed, but in letting Crusoe re-invent all the branches of the arts of agriculture and manufacturing he fictionalised these ideas, creating in the microcosm of Crusoe's island a complete cycle of human activities. In effect, *Crusoe* is Defoe's 'living' History of Trade.

Necessity does not hinder Crusoe, rather it inspires him to be resourceful and inventive. 'As it is proverbially said, that *necessity is the mother of inventions*, so experience daily shews, that the want of subsistence, or of tools and accommodations, makes crafts-men very industrious and inventive, and puts them upon employing such things to serve their present turns, as nothing but necessity would have made even a knowing man to have thought on' (my italics). Although containing Defoe's favourite proverb, that sentence is not his but Boyle's, in his *Usefulness of Experimental Natural Philosophy* (1663–71), that is, in Boyle's appeal for a History of Trades (Boyle 1744, III: 168). One of the many instances where Defoe employs the proverb is in his *History of Trade* where he writes: 'Necessity which is the Mother, and Convenience which is the Handmaid of Invention, first Directed Mankind from these Originals, to Contrive Supplies and Support of Life. Corn was not to be Eaten whole, but receiv'd, lay'd up, brought to maturity, and then suffer due preparations to make it the better become Food proper for the support of Human Life' (I: 31).[7] The proverb does not appear in *Crusoe*, but adaptations of it occur with such frequency that they become imprinted on the reader's mind. On 4 November 1659, in one of his first entries into his journal, Crusoe notes that 'Time and Necessity made me a compleat natural Mechanick soon after' (p. 72). Reviewing his first decade on the island he reports: 'I improv'd my self in this time in all the mechanick Exercises which my Necessities put me upon applying my self to' (p. 144).[8] Necessity gives the impulse to Crusoe's industry and improvements; practical knowledge of mechanical skills is the direct result. Crusoe can be recognised as the craftsman or amateur scientist described by Boyle, who, seized with the zeal for experimenting, turns his position of 'Extremity' to his advantage.

Other reflections are pertinent here. When the Fellows of the Royal Society composed their *'Directions for Sea-men going into the East & West-*

---

[7] In his *Review* for 21 June 1711 Defoe turns these ideas into an allegory, making Necessity the illegitimate daughter of Pride and Sloth (VIII: 153–6).
[8] For further references see pp. 49, 68–9, 73–4, 77, 104–5, 107, 117–18, 119–23, 134–5.

Indies, the better to capacitate them for making such observations abroad, as may be pertinent and suitable for their purpose', they instructed as follows:

> To observe the Declinaton of the *Compass*, or its Variation from the *Meridian* of the place, frequently; marking withal, the *Latitude* and *Longitude* of the place, wherever such Observation is made, as exactly as may be, and setting down the *Method*, by which they made them.
>
> To carry *Dipping Needles* with them, and observe the Inclination of the Needle in like manner.
>
> To remark carefully the Ebbings and Flowings of the Sea ...
>
> To keep a Register of all changes of Wind and Weather at all houres, by night and by day ...
>
> To observe and record all Extraordinary Meteors, Lightnings, Thunders, *Ignes fatui*, Comets, etc. marking still the places and times of their appearing, continuance, etc.[9]

In addition to these 'Directions' the traveller was advised to keep an 'exact *Diary*' of everything worth remembering. The traveller should

> always have a Table-Book at hand to set down every thing worth remembring, and then at night more methodically transcribe the Notes they have taken in the day ... Every Traveller ought to carry about him several sorts of Measures, to take the Dimensions of such things as require it; a Watch by which, and the Pace he travels, he may give some guess at the distances of Places ... a Prospective-glass, or rather a great one and a less, to take views of Objects at greater and less distances; a small Sea-Compass or Needle, to observe the situation of Places, and a parcel of the best Maps to make curious Remarks of their exactness, and note down where they are faulty.[10]

Reading the Society's 'Directions' and being now aware of Defoe's closeness to the Baconian habit of mind, we wonder whether it is by chance that Crusoe is allowed to begin his island existence furnished with 'three or four Compasses, some Mathematical Instruments, Dials, Perspectives, Charts, and Books of Navigation' (p. 64). It appears that Crusoe is from the start equipped for making the precise scientific observations demanded by the Fellows. Equally, we can now see Crusoe's registers of the weather within the context of the activities of

---

[9] *Philosophical Transactions* 1665–7, I: 141–3.

[10] From the anonymous 'Introductory Discourse, containing The whole History of Navigation from its Original to this Time', in Churchill 1704, I: lxxv–lxxvi, where this advice is preceded by a reprint of the Royal Society's 'Directions' for travellers.

the Royal Society. As we have seen, the members of the Society explicitly instructed travellers 'to keep a Register of all changes of Wind and Weather'. Hooke, Boyle, Locke, Christopher Wren, Wilkins were among those engaged in compiling histories of the weather. That Defoe's teacher also taught the value of making a history of the weather has already been mentioned (see pp. 30, 70 above, and p. 136 below). Here, to compare with Crusoe's register of the weather, is an excerpt from Robert Hooke's 'Method for making a History of the Weather' (see fig. 2) which was printed in the *History of the Royal Society* (Sprat 1959: 173–9); for Crusoe's weather-chart see *Crusoe* 1: 106). I would suggest that just as it is no coincidence that the 'trial and error' method of the Baconian experimenter should be strikingly characteristic of Crusoe's 'improve[ment] in all mechanick Exercises', so it is not by chance that he should begin his island sojourn furnished with scientific instruments, charts and maps, that he studies the weather and has 'a Table-Book at hand to set down every thing worth remembering'. What connects Crusoe with the Royal Society is that he fulfils, not in one but in a whole range of his activities, the Fellows' 'Directions' for gaining and recording accurate knowledge of natural phenomena.

Since the publication of both G. A. Starr's and J. P. Hunter's critical evaluations of *Crusoe* in the 1960s, Crusoe's journal has been related to Defoe's religious background (see n. 15 below). Crusoe's diary, they argued, was inspired by the Puritans' (and other sects') demand to keep a record of one's own progress (or relapse) towards salvation. I am not denying the influence of popular religious beliefs and the Dissenters' call for spiritual book-keeping may well be one of the inspirations for Crusoe's journal. This does not, however, rule out the fact that a very different tradition of thought also contributed to the complexity and richness of Crusoe. That the Society's directions for making faithful reports of every thing and notion left its impress on the novel has been, and will be further, demonstrated; Crusoe's habit of logging every minute particularity of his experience is part of this tradition.[11] Crusoe himself is, of course, unaware of the Society's guidelines, but there can be little doubt that Defoe consciously followed their general rules for

---

[11] J. P. Hunter was right to comment that '*Robinson Crusoe* ultimately is much more complex than any of the traditions which nourish it, but the complexity should not obscure the ancestry' (1966: 50).

The *Form* of a *Scheme*.

Which at one view represents to the Eye Observations of the Weather, for a whole Month, may be such, as follows.

Days of the Moneth, and Place of the Sun	Remarkable hours.	Age and Sign of the Moon at Noon.	The Quarters of the Wind, and its strength.	The Faces or visible appearances of the Sky.	The Notablest Effects	General Deductions. These are to be made after the side is filled with Observations, as
June 14 ♊ 12.46′	4 8 12 4 8 12	27 ♉ 9. 46 Perigeum	W----2 ------3 ------3½ ------- WSW 1 -------	Clearblue, but yellowish in the N E. Clouded toward the South. Checkered blue.	A great Dew Thunder far to the S. A very great Tyde.	From the last Quarter of the Moon to the Change, the weather was very temperate, but for the Season, cold ; the Wind pretty constant between N. and W. &c.
15 ♊ 13.40′	8 4 6 12	28 ♉ 24.51	NW 3 4 N 2 1	A clear sky all day, but a little checker'd about 4 P. M. At Sun-set red and hazy.	Not by much so big a Tyde as yesterday. A great Thunder-Showre from the N.	
16. ♊ 14. 57 &c.	10	New Moon at 7. 25. A. M. ♊ 10.8 &c.	S 1 &c.	Overcast and very lowring, &c.	No dew upon the ground, but very much upon Marblestones, &c.	

2 Robert Hooke's 'Scheme representing at one View to the Eye the Observations of the Weather for a Month' from *The Philosophical Transactions of the Royal Society*.

making objective and credit-worthy observations. The awareness is Defoe's. He is convinced that the 'effectual, and unanswerable Arguments' of facts will persuade the reader of the authenticity of Crusoe's account; it is Defoe who in the disguise of 'the Editor' compiled a *'just History of Fact'* (Preface).

It is worth recalling at this point that Defoe makes Crusoe begin his island experience in September 1659, that is, in the year leading up to the foundation of the Royal Society. Thus at the same time as we imagine Crusoe on his remote island studying 'every mechanick Art', the Fellows, and those enlisted to help, were doing exactly the same. While Crusoe experiments and then registers his experiments of sowing and growing of corn and baking of bread, the Fellows discussed and recorded agricultural and mechanical improvements. It was noted on 25 March 1663 that 'Dr. Wilkins gave an account of the way and benefit of setting corn by a peculiar engine; and was desired to send for it from Oxford, and to communicate to the society in writing the substance of what he had related to them this day concerning the matter' (Birch 1756, I: 213). On 1 March 1665 Evelyn read his history of bread-making to the Society, while another meeting recorded that Boyle had received an account from his gardener about the sowing and growing of potatoes (see p. 87 above and p. 126 below).[12] Indeed, as has been shown already, all of Crusoe's activities during his solitary stay on the island are part of the 'compleat Cycle and History of Trades' as compiled by the Fellows of the Society.

Much of Crusoe's time is taken up with establishing order in his environment: 'I had every thing so ready at my Hand, that it was a great Pleasure to me to see all my Goods in such Order' (p. 69). Next to his well-ordered 'Magazine' of provisions and tools, Crusoe eventually has a 'living Magazine of Flesh, Milk, Butter and Cheese' (p. 153). He writes of his boat – though he might have said it of everything about him – that he 'kept all Things about or belonging to [him] in very good Order'. He tames some wild goats, and builds an enclosure in order to keep 'the tame from the wild' (p. 146). This

---

[12] Evelyn's history of bread-making, his *Panificium, or the Several Manners of Making Bread in France*, was published in John Houghton 1681–3; Evelyn's history was a direct response to Bacon's recommendation for a 'History of Baking, and the Making of Bread'.

For Petty's invention of an agricultural 'engine' which could sow seed automatically see C. Webster 1968–9: 367.

idea, the separation and protection of culture from nature, or, as he puts it, keeping the tame from 'always running wild', is behind Crusoe's concern with order. He registers and divides the seasons of the year; he tells us that he 'began to order my times of Work ... time of Sleep, and time of Diversion' (p. 72). Similarly, 'like Debtor and Creditor', he separates and tabulates the evil and good aspects of his situation (pp. 65–6). Crusoe enforces order, imposes his values upon his surroundings. Ultimately, the kingdom over which he presides is a huge, tidy 'magazine' of things and notions. The Baconians referred to their histories or taxonomies as 'magazines' or 'storehouses' of matter and notions (Bacon 1857, IV: 255). In the manner of these historians of nature, Crusoe compiles, divides and ranks his collected data. Item by item, 'Piece by Piece' as he informs us, he assembles and classifies a complete world. He taxonomises, however, not for definition but for use. Unlike the Baconian natural taxonomies, which set out to register a supposedly finite number of things into a preconceived order, Crusoe lives to discover the order, value and usefulness of things.

Significantly, Crusoe's empire over things is already declared in his tenth month on the island: 'I was King and Lord of all this Country indefeasibly, and had a Right of Possession; and if I could convey it, I might have it in Inheritance, as compleatly as any Lord of a Mannor in *England*' (p. 100). In the fourth year, he considers himself 'Lord of the whole Mannor; or if I pleas'd, I might call my self King, or Emperor over the whole Country which I had Possession of. There were no Rivals' (p. 128). Two decades later, Crusoe repeats with little variation: 'There was my Majesty the Prince and Lord of the whole Island; I had the Lives of all my Subjects at my absolute Command ... and no Rebels among all my Subjects' (p. 148). Traditional interpretations see Crusoe's mastery as a gradually evolving theme. But the plain fact is that his dominion is established before the end of the first year on the island. The belief in man's knowledge of, and consequent power over things, is a granted – perhaps we should say – an inherited maxim: it is the foundation upon which Crusoe's preoccupation with order rests. Crusoe not so much learns as reflects and applies what Bacon's philosophy of things had taught. The Fellows of the Royal Society had recommended men to 'rank all the *varieties*, and *degrees* of things, so orderly one upon another; that standing on the top of them, we may perfectly behold all that are below, and make them all serviceable to the quiet, and peace, and plenty of Man's life' (Sprat 1959: 110). What is gradually and increasingly made clear in the first half of the book is that by ordering everything, Crusoe succeeds in making all things serviceable to the quiet, peace and plenty of his life. The scene depicted immediately before he encounters the single footprint is one of absolute peace founded upon absolute power over things. It is at the zenith of his unchallenged dominion that his peace of mind is shattered; knowledge of, and power over nature are put into question as we come to 'a new Scene of [his] Life' (p. 153).

Nature, man and God

For the sake of clarity, the focus has so far been on Crusoe's concept of the world of nature and of man. It would, however, be wrong to detach his empirical investigations from his religious preoccupation, for clearly the two go together, the search for the knowledge of things assisting the knowledge of religious principles.

In his reconciliation of part-secular and part-Christian ideals, Defoe testifies to a body of beliefs generally known as natural theology. Very briefly its principles are these. Thanks to the wisdom of God, the universe is harmoniously created for man's benefit and use. There is order and purpose in all things, even random occurrences such as natural disasters are essential to carry out God's will to man. Contemplating nature, man is filled with wonder at the divine order; consequently, it is right that he should study nature. By following the footsteps of nature, man can discern the will of God – that is, Providence is revealed by science. Yet, finally, science has its limitations: man is granted to know only aspects of the divine creation, God alone has perfect knowledge of the essences of things.

The point to be stressed is that natural theology transcended theological and political boundaries. Robert Boyle, who perhaps more than anyone else encouraged the reconciliation of scientific and religious pursuits, was a convinced member of the Church of England (see p. 27 above). In this respect, Boyle was in agreement with the Baconian Puritan reformer, John Dury, who declared in *A Seasonable*

Discourse (London, 1646): 'What art or science doth not advantage mankind, either to bring him nearer unto God in his soul, or to free him from the bondage of corruption in his body, is not at all to be entertained; because at best it is but a diversion of the mind' (in Hill 1965: 85). By comparison John Wilkins (latitudinarian and later Anglican bishop) wrote: 'Our best and most divine knowledge is intended for action, and those may justly be counted barren studies, which doe not conduce to practise as their proper end.' Wilkins, who is often considered to epitomise the fusion of Puritan and Baconian aims, opened *Ecclesiastes; or, A Discourse concerning the Gift of Preaching* with the observation that 'the end of all Sciences and Arts [is] to direct men by certain rules ... in their knowledge and practise' to God (Wilkins 1648: 3 and Wilkins 1646, respectively). Wilkins had not the slightest doubt that science could prove God and a Providence.[13] Although differing on many points of doctrine from Wilkins, the Dissenter Charles Morton shared his belief that 'the End and last design of the science it self is to enable a man to contemplate' the mystery of the creation (Morton 1940, Preface).

What united the defenders of natural theology was their commitment to the progress of New Science. Irrespective of their denominational background, they believed that exact observations of natural phenomena would provide evidence of the existence of God. Sprat put it well when he observed that, in a time of political and theological unrest, experimental science offered a pursuit in which 'the *Soldier*, the *Tradesman*, the *Merchant*, the *Scholar*, the *Gentleman*, the *Courtier*, the *Divine*, the *Presbyterian*, the *Papist*, the *Independent*, and those of *Orthodox Judgment*, have laid aside their names of distinction, and calmly conspir'd in a mutual agreement of *labors* and *desires*' (Sprat 1959: 427). I am emphasising this point because some interpreters of *Crusoe* have argued as if only the Puritans had seen mundane events charged with moral and spiritual meaning (for example, J. P. Hunter 1966 and Starr 1965; and see p. 119 below).

The context into which Crusoe fits most naturally is that of the Christian virtuosi. Their argument had been that by studying 'the wonderful Order, Law and Power' of nature, man could discern the

---

[13] Cf. Wilkins 1640: 237–40. On Wilkins's personification of the fusion of Puritan and Baconian aims see Hill 1965: 130.

Creator's 'wonderful Effects'. 'I say wonderful', Hooke went on, because 'every natural Production may be truly said to be a Wonder or Miracle'. While the 'observing Naturalist may perhaps tell the Steps or Degrees he has taken notice of in its Progress from the Seed to the Seed', and 'may also tell the Times and Seasons in which these Progresses have been or will be performed', he cannot know 'the moving Power ... there is the Miracle that he may truly admire but cannot understand'. It is worth noting that Hooke's reflections appeared in his 'Discourse of Earthquakes' (delivered 23 July 1690), where he argued – just as Crusoe is made to argue – that natural disasters are God's way of revealing himself to man (Hooke 1705: 423–4).

Crusoe's observations of nature reveal to him two things: first, he gains practical knowledge and learns all the 'trades' on which his physical survival depends; second, he discovers that there is an 'invisible Power which alone directs such Things' (p. 90). His readiness to 'acquiesce in the Dispositions of Providence' assures his spiritual survival. Nothing illustrates the unfolding of Crusoe's awareness of a 'Supreme Being' better than the episode of the accidental crop. At first, when his carelessly discarded seeds of corn begin to grow, Crusoe imagines 'that God had miraculously caus'd this Grain to grow without any Help of Seed sown' (p. 78). But recollecting that it was he himself who had carelessly thrown the husks of corn away, the miracle shrinks into an ordinary event, and 'the wonder began to cease'. After two years of diligently observing and experimenting, he succeeds in growing a crop and he considers himself 'Master of my Business'. It is only after the fourth year that Crusoe has acquired the correct and 'different Knowledge from what [he] had before' which lets him see that 'nothing but a Croud of Wonders could have' made his daily bread grow. Now he reflects with gratitude that he is being fed 'by a long Series of Miracles', and so 'gives daily Thanks for that daily Bread' – the daily bread, as in Our Lord's Prayer, eventually standing for all the needs supplied by God (p. 132).

An exact parallel to Crusoe's growing consciousness of an omnipotent power directing all things, can be found in John Ray's *Wisdom of God manifested in the Works of the Creation* (London, 1691). Defending his belief in a deity as a foundation of all things, Ray, the quintessential Baconian natural historian, instanced even

> *illiterate Persons ... affirming, that they need no Proof of the being of a God, for that every Pile of Grass, or Ear of Corn, sufficiently proves that. For, say they, All the men of the World cannot make such a thing as one of these; and if they cannot do it, who can, or did make it but God? To tell them that it made it self, or sprung up by chance, would be as ridiculous as to tell the greatest Philosopher so.*
>
> (Preface)

Crusoe, the 'illiterate', untaught person described by Ray, momentarily believes that the seeds have 'sprung up by chance'; it is only later, after he has gathered 'new Knowledge' that he finds in the 'Ear of Corn' sufficient proof that 'nothing but a Croud of Wonders' could have made it. A clearer statement of Crusoe's (or rather Defoe's) alignment with the Christian scientists' belief that the wisdom of God is made manifest in the creation, could hardly be found.

Although part 3 of *Crusoe* has, since the comments of Charles Gildon, frequently been regarded (or even discarded) as an unrelated afterthought, a belated defence of the autonomy of the original volume, it is in this volume that we have the most direct explanation for Crusoe's meticulous observation of reality. In the chapter entitled 'Of Listening to the Voice of Providence' Defoe has Crusoe point out that

> To listen to the voice of Providence, is to take strict notice of all the remarkable steps of Providence which relate to us in particular, to observe if there is nothing in them instructing to our conduct, no warning to us for avoiding some danger, no direction for the taking some particular steps for our safety or advantage, no hint to remind us of such and such things omitted, no conviction of something committed, no vindictive step, by way of retaliation, marking out the crime in the punishment.
>
> (p. 205)

That Crusoe here interprets his habit of 'tak[ing] strict notice of all the remarkable steps', dates and events during his island experience, is beyond question. For the present purpose of investigating Defoe's alignment with the principles of experimental science, it is significant that Crusoe reveals an attitude characteristic of the Christian virtuosi. Recognising a relation between science and religion, they believed that by observing 'such conjunctures of circumstances' man might get insight into 'the genuine consequences of the order [God] was pleased to settle in the world' (Boyle 1744, IV: 362). Nature carries the imprint of the Creator, and by studying her rightly she may give moral and religious instructions for future conduct.

In the same chapter, Defoe has Crusoe refer to Solomon, the prototype of the Baconian searcher into nature. Explaining that systematic investigations assist us to be good Christians, Crusoe quotes from Proverbs 2: 4, where Solomon 'bid us cry after knowledge, ... dig for her as for silver, and search for her as for hid treasure. It is certain here that he meant religious knowledge.' While Solomon is unrelenting in his quest for knowledge, the fool 'sits down in his ignorance, repulsed with imaginary difficulties, without making one step in the search after the knowledge which he ought to dig for as for hid treasure' (p. 176). That Crusoe is to be identified with Solomon and not with the fool is so obvious that it hardly needs stating. Nothing could be further from Crusoe's habit of mind than 'to sit down in his ignorance, repulsed with imaginary difficulties'. A few years later, when Defoe returns to this subject in his *History of Arts and Sciences*, he contrasts Solomon's fool with the experimental scientists, who,

> having open'd a Door into the vast Ocean of Mathematical Knowledge, it fir'd their Souls with a happy desire of knowing more; I say fir'd, because Mankind has ever since had an unquenchible Thirst after the compleat Discovery of Nature, and the highest degree of acquir'd Knowledge, and an indefatigable Application to farther and farther Improvements in Arts and Science; in a word, in all possible Degrees of Learning and Knowledge.
>
> (p. 232)

This is one of those many instances where Defoe quotes himself, and it is noteworthy that the first time he had used these words was in his description of Crusoe's conscientious search into nature. In 'Robinson Crusoe's Preface' to part 3 we read: 'Here is invincible patience recommended under the worst of misery, indefatigable application and undaunted resolution under the greatest and most discouraging circumstances' (pp. xii–xiii). Crusoe is cast in the role of Solomon, Bacon's example to all future ages of systematic, scientific exploration of nature. Following Solomon, Crusoe is shown to 'dig his knowledge out of the hard mines of experience'; he epitomises the experimental scientists' 'unquenchible Thirst after the compleat Discovery of Nature' (cf. Bacon 1857, III: 219 and see pp. 15, 60 above and pp. 175–6 below).

The work in which Defoe first explicitly declared his adherence to the principles of Baconian science was *The Storm* (1704). Defining the experimental philosopher's activity, he wrote that 'it is not enough for him to know that God has made the heavens, the moon, and the stars, but must inform himself where he has placed them, and why there; and what their business, what their influences, their functions, and the end of their being' (p. 262). It seems to have gone unnoticed so far that the text Defoe had in mind when he wrote this defence of New Science was Psalm 8:

> 3 When I consider thy heavens, the work of thy fingers, the moon and the stars which thou hast ordained;
> 4 What *is* man, that thou art mindfull of him? and the son of man, that thou visitest him?
> 5 For thou hast made him a little lower then the angels, & hast crowned him with glory and honour.
> 6 Thou madest him to have dominion over the works of thy hands; & thou hast put all *things* under his feet.[14]

Verse 3, present in Defoe's mind when he defended the experimental scientist's activity in *The Storm*, reappears in both parts 1 and 3 of *Crusoe*. The direct reference comes in the final part, where Crusoe is made to argue that 'the voice of God [is heard] in His works', and Crusoe quotes: 'When I view the heavens, the work of Thy hands, the moon and the stars which Thou hast made, then I say, what is man?' Crusoe goes on to reason that, as the works of the Creator 'fill us with wonder and astonishment, admiration and adoration ... it is without question our wisdom and advantage to study and know them, and to listen to the voice of God in them' (p. 195). Crusoe's reflection strongly calls to mind the following passage from Ray's *Wisdom*:

> And to me it seems, that where the Heavens and Earth, and Sun, and Moon, and Stars, and all other Creatures are called upon to Praise the Lord; the meaning and intention is, to invite and stir up Man to take notice of all those Creatures, and to Admire and Praise the Power, Wisdom and Goodness of God manifested in the Creation and Designations of them.
>
> (p. 132)

---

[14] *The Whole Book of Psalms collected into English Metre*, T. Sternhold, J. Hopkins and others (Cambridge, 1661).

Bearing in mind Defoe's use of Psalm 8 in his defence of experimental science in *The Storm*, and his repetition of his argument in *Crusoe* 3, we can see that this text once more underlies Crusoe's question in the first volume:

> What is this Earth and Sea of which I have seen so much, whence is it produc'd, and what am I, and all the other Creatures, wild and tame, humane and brutal, whence are we?
> Sure we are all made by some secret Power, who form'd the Earth and Sea, the Air and Sky; and who is that?
>
> (p. 92)

Parallels between the works have been noted before; to these I would like to add the above (see Secord 1924: 78–85, Rogers 1979: 57–8). Although fifteen years (and according to Moore's *Checklist* more than 300 works) intervened between *The Storm* and parts 1 and 3 of *Crusoe*, it is clear that Defoe recalled the earlier work when he composed *Crusoe*. He uses Psalm 8 (the text, incidentally, with which Ray concluded the Preface of his *Wisdom*) to justify Crusoe's scientific search into nature. Crusoe, in Defoe's words from *The Storm*, is the Christian virtuoso who 'search[es] the steps [nature] takes, the tools she works by; and, in short, [comes to] know all that the God of nature has permitted to be capable of demonstration' (p. 262). Following this method, Crusoe reaches the conclusion that if God has made all things, then he must also have the power to guide and govern them all, 'for the Power that could make all Things, must certainly have Power to guide and direct them' (*Crusoe* 1: 92).

Man's empire over things is the theme of *Crusoe*, and woven into it is another, namely, Crusoe's increasing awareness of his subservience to God. When God created the world, he ordained that all things be subjected to the dominion of man; yet, it was also the divine will that man govern not as absolute king but 'as viceroy to the King of all the earth'. This is the lesson that Crusoe has to learn. Originally he thought of himself as 'King and Lord of all this Country indefeasibly ... and if I could convey it, I might have it in Inheritance, as compleately as any Lord of a Mannor in *England*' (p. 100). But the wiser and more devout Crusoe is aware that 'it cannot be conceived, without great inconsistency of thought, that this world is left entirely to man's conduct, without the supervising influence and the secret direction of the Creator'. It is within the context of his definition of natural religion that Defoe has Crusoe explain that while the earth is given to man as 'an inheritance' and 'subjected to his authority', man is *not* the owner but the 'tenant to the great Proprietor, who is Lord of the manor, or Landlord of the soil' (*Crusoe* 3: 179). Direct comparisons are always useful for making a point, and by re-using the same words in parts one and three of the work Defoe highlights Crusoe's changed position.

By attending the 'Dictates and Directions' of Providence, Crusoe becomes aware that it 'was my unquestion'd Duty to resign my self absolutely and entirely to [God's] Will' (*Crusoe* 1: 157). To subject himself to God's will means to re-organise the rhythm of his daily life. Initially, he had divided his day into periods of work, recreation and rest, but after three years he replaces this division with one that not only incorporates daily readings of the Bible, but asserts the supremacy of religion by putting it head of the list: '*First*, My Duty to God, and the Reading the Scriptures ... *Secondly*, The going Abroad with my Gun for Food ... *Thirdly*, The ordering, curing, preserving and cooking what I had kill'd or catch'd for my Supply' (p. 114). Equally, in his division of the week: 'I had all this Time observ'd no Sabbath-Day; for as at first I had no Sense of Religion upon my Mind'; but now he decides to divide his week into secular and religious time (pp. 103–4).

By taking 'strict notice of all the remarkable steps' of nature Crusoe learns that even those things which appear at first to thwart our intentions are performed for our best use and means (cf. Wilkins 1649: 5, 11). Thus Crusoe comes to interpret the earthquake and his illness as God's way of admonishing him and convincing him of the wisdom of revelation. Crusoe (or rather Defoe) is at one with those Baconians who declared that the concurrence of events tends 'to the illustration of God's wisdom, to have so framed things at first, that there can seldom or never need any extraordinary interposition of his power' (Boyle 1744, IV: 361–2). To see Crusoe's veneration of the 'Dictates of Nature' as a specifically Puritan phenomenon is too narrow an interpretation. Such an interpretation leaves out, or undervalues, the scientific aspects, that is, such aspects as observation and collection of precise data, exploration and experiments, which give meaning to Crusoe's spiritual development.[15]

Crusoe learns his subjection to 'the great Governour of all Things'; in turn, the world around him acknowledges his mastery and the fact that, in the words of the Psalmist, 'all *things* [are] put under his feet'. This is graphically illustrated in Friday's symbolic gesture when he laid 'his Head upon the Ground' and 'taking [Crusoe] by the Foot, set [his] Foot upon his Head' (pp. 203–4). Later, the scene is repeated and its meaning made even more explicit:

> At last he lays his Head flat upon the Ground, close to my Foot, and sets my other Foot upon his Head, as he had done before; and after this, made all the Signs to me of Subjection, Servitude, and Submission imaginable, to let me know, how he would serve me as long as he liv'd.
>
> (p. 206)

Just as Friday accepts his subjection, so Crusoe takes his superiority for granted.

Different as Defoe and Crusoe are in many ways, in their attitude to the savage they are united. Defoe's most commonly held view was that in the hierarchy of the great chain of being, the savage was placed below the European, and only slightly above the animals. Deprived of the civilising effects of education and Christian faith, the savage was considered inferior, blind, ignorant, 'brutish' and 'barbarous' (p. 217). Natural man is 'a plain coarse Piece of Work', but 'Nature and Art joyn'd, make an exquisite and accomplish'd Piece' of him.[16] Left in the 'state of mere Nature' into which he was born, the savage continues a pre-logical, childish, or, less, an animal existence. What distinguishes us 'from Brutes', Defoe declares, is the 'rational soul', and 'education carries on the distinction and makes some less brutish than others. This is too evident to need any demonstration' (*Projects*, p. 145). Defoe judges that 'an untaught Man', a 'Creature in human Shape, but intirely neglected and uninstructed, is ten thousand times more miser-

---

[15] J. P. Hunter and G. A. Starr were among the first who tried to give the spiritual element in the novel its due; in their effort to redress the balance, it seems to me, they over-emphasised the Puritan aspect: J. P. Hunter 1966, Starr 1965: 74–125. For a discussion that takes into account Defoe's Puritan upbringing and his empiricist rationalism see Stamm 1936, and see H. Fisch 1952, McAdoo 1965, and Downie 1983.

[16] See *Mere Nature Delineated*, p. 68 and *Present State*, pp. 300–1.

able than a Brute'. While education civilises and polishes a man, deprivation of such refinement stunts both the body and the soul (*Mere Nature Delineated*, p. 63 and cf. *Chickens Feed Capons* (1731), p. 5). Like nature herself, 'primitive', less-developed cultures welcome education and improvement. In his *Historical Account of Sir Walter Raleigh* Defoe records how the inhabitants of Guinea invited English colonisation, how country and people waited to be 'Possess'd, Planted and Secur'd' by Western civilisation. In his portrayal of Crusoe's relationship with Friday Defoe defends the most generally accepted view of the seventeenth and early eighteenth centuries. It was later, in the 'age of enlightenment', that the savage was ever thought of as 'noble'; it is only in this century that social anthropologists have come to see and respect the fundamentally different yet complex customs and beliefs of the so-called 'primitives' (Beattie 1964: 65ff).

According to Crusoe's thinking, God has bestowed upon Friday 'the same Powers, the same Reason, the same Affections, the same Sentiments of Kindness and Obligation, the same Passions and Resentments of Wrongs, the same Sense of Gratitude, Sincerity, Fidelity, and all the Capacities of doing Good, and receiving Good, that he has given to us' (p. 209). What separates Crusoe from Friday is education and spiritual guidance, 'the great Lamp of Instruction'. Crusoe records that he 'made it [his] Business to teach him every Thing, that was proper to make him useful, handy, and helpful; but especially to make him speak, and understand me when I spake' (p. 210). And Friday, recognising Crusoe's cultivating effect, observes: '*You do great deal much good ... you teach wild Mans be good sober tame Mans; you tell them know God, pray God, and live new Life*' (p. 226). Like Adam in the Garden of Eden, Crusoe orders, controls and names wild nature: 'First, I made him know his Name should be *Friday* ... I likewise taught him to say Master, and then let him know, that was to be my Name' (p. 206). Years of experience are now freely shared with Friday; ripe for knowledge, Friday learns the manufacturing and agricultural crafts within a few months. Not only does he adopt European customs, eat salt, wear clothes, learn how to use a gun, but he accepts from Crusoe the basic principles of Christian faith.

The belief that knowledge should be made readily available to all is one of the distinguishing features of modern, that is, Baconian, science.

Scientific progress depended on the moral conviction that knowledge should be shared; this belief was reinforced by Christian doctrine, which also urged that all things be shared for the benefit and help of our fellow men. Defoe's devotion to the promotion of useful knowledge for the good of all has been fully demonstrated. As we have seen, he was convinced that 'Science, being a publick blessing to mankind, ought to be extended and made as difusiv as possible, and should, as the Scripture sayes of sacred knowledge, spread over the whole earth, as the waters cover the sea' (*Gentleman*, pp. 197–8). In Crusoe's eagerness to teach and share his knowledge, and Friday's willingness to learn, is fulfilled the first condition for the advancement of modern science.

Language and function

Any discussion of *Crusoe* demands consideration of Defoe's prose style in general. Again, part 3 helps to direct us. Equating plainness with honesty, Defoe writes: 'The plainness I profess, both in style and method, seems to me to have some suitable analogy to the subject, honesty, and therefore is absolutely necessary to be strictly followed.' It is for this reason that he chose 'a natural infirmity of homely plain writing'. Defoe goes on to argue that 'the plainness of expression, which I am condemned to, will give no disadvantage to my subject, since honesty shows the most beautiful, and the more like honesty, when artifice is dismissed, and she is honestly seen by her own light only' (*Crusoe* 3, p. 23). Sixteen years earlier Defoe had drawn the same analogy in *The Storm*, where he declared that: 'The plainness and honesty of the story will plead for the meanness of the style in many of the letters ... These come dressed in their own words because ... I am persuaded, they are all dressed in the desirable, though unfashionable garb of truth' (pp. 257–8). Contrasting honest plainness with deceitful rhetoric, Defoe invariably insists that the 'plain and homely style' is the perfect style:

> easy, plain, and familiar language is the beauty of Speech in general, and is the excellency of all writing, on whatever subject, or to whatever persons they are that we write or speak. The end of Speech is that men might understand one another's meaning; certainly that speech, or that way of speaking which is most easily understood, is

the best. If any man was to ask me, what I would suppose to be a perfect stile or language, I would answer, that in which a man speaking to five hundred people, of all common and various capacities, Ideots and Lunaticks excepted, should be understood by them all, in the same manner with one another, and in the same sense which the speaker intended to be understood, this would certainly be a most perfect stile.

(*Tradesman*, p. 26)

The question that naturally arises within the context of the present investigation of Defoe's indebtedness to the New Sciences is whether Defoe's preference for 'a natural infirmity of homely plain writing' was influenced by the Royal Society's preference for 'the language of Artizans, Countrymen, and Merchants, before that, of Wits, or Scholars' (Sprat 1959: 113)?

Ian Watt was the first to recognise a similarity between Defoe's prose style and that recommended by the Society. Watt quotes from the famous passage in *The History of the Royal Society* where Sprat sets down the Fellows' resolution to reject stylistic embellishments and return to 'a close, naked, natural way of speaking; positive expressions; clear senses; a native easiness: bringing all things as near the Mathematical plainness, as they can' (p. 113). 'Certainly,' writes Watt, 'Defoe's prose fully exemplifies the celebrated programme of Bishop Sprat.' However, this acknowledgement is followed by a denial of any direct connection, since 'Defoe naturally preferred such language [as] he was a merchant himself'. According to Watt, Defoe's evident 'mathematical plainness' is 'suited to carrying out the purpose of language as Locke had defined it, "to convey the knowledge of things"' (Watt 1972: 113–14).

Taking up Watt's point in 1964, M. Novak began by citing a passage from the *Review* in which Defoe discussed whether in order to convey the idea of a thing, it is better to describe 'the Thing it self' or to use 'Emblems and Figures', that is, metaphor (the passage is quoted on pp. 60–1 above). At first Novak conceded that Defoe might have derived his ideas from 'Bacon or any number of writers', but then Defoe's reference to the 'Doctrine of Ideas', Novak argued, suggests Locke's direct influence (Novak 1964: 661). As I have tried to show, Defoe's argument for describing 'the Thing it self' is remarkably close to Bacon's. Furthermore, Locke was one of Bacon's most devoted disciples, his theory of language being directly traceable to Bacon's philosophy of things.[17]

One year later, in 1965, James Boulton reaffirmed that 'the "Mathematical plainness" of language celebrated by Bishop Sprat was [Defoe's] proper medium' (Boulton 1975: 3). Again, Robert Adolph, in *The Rise of Modern Prose Style* (Cambridge, Mass., 1968) relates Defoe's style to 'The New Prose of Utility' put forward by the Baconian scientists. Adolph, however, overstates both the Society's and Defoe's concern with utility. The standard of prose style which the Society recommended was determined by *all* the tenets of its philosophy, not just by its dedication to usefulness (see pp. 42–3 above). G. A. Starr confronted the question of Defoe's indebtedness to the Royal Society squarely. He was the first to recognise, within this context, that the Fellows claimed to eschew rhetoric and metaphor yet used these stylistic devices in order to make a point clearly and easily. Starr's conclusion was that Defoe was indeed preoccupied with the 'world of things' but that he was 'less concerned with rendering external things directly than with presenting them as experienced by or related to his narrators' (Starr 1974: 293–4). The study that most specifically drew a parallel between the concern and prose style of the *Philosophical Transactions* and *Crusoe* appeared in 1984 by M. Baridon and was entitled 'Le style de Defoe et l'épistémologie de la "New Science"' (Baridon 1984 and see F. H. Ellis 1985). Similarly, Melinda Snow, in 'The origins of Defoe's first-person narrative technique: an overlooked aspect of the rise of the novel', argued for the Society's direct influence on Defoe's plain mode of communication (Snow 1976).

Generally speaking, it seems that Defoe scholars agree about certain aspects of his style. Critics have acknowledged that he prefers 'a native easiness' and that there exists a predominance of words of Anglo-Saxon origin; he writes naturally, directly and plainly, stressing the knowledge of things rather than of words; he rejects the 'educated', studied language of scholars. It appears further that while scholars see a resemblance or even indebtedness to the stylistic standards of the Royal Society, they have been reluctant to admit an immediate link with the Baconian philosophy.

[17] See pp. 60–1 above. For references arguing for Locke's indebtedness to Bacon see Anderson 1948: 299, 302, Aarsleff 1964: 165–88 and Hill 1965: 126.

Several reasons account for this phenomenon. Of the explanations to be suggested, the first and most important is that, so far, neither the critics of Defoe nor the historians of science have recognised Morton's extensive knowledge and use of the Baconian scientists. In the light of the discussion offered in chapters 3 and 4 we are now aware of Morton's indisputable familiarity with a great variety of the Royal Society's activities. Defoe's teacher knew of the relation between science and language. We have seen that when Morton gave advice on sermon-writing he turned to John Wilkins, the man most influential in shaping the Society's attitude to prose. Not only did Morton recommend Wilkins's popular handbook to plain preaching, *Ecclesiastes*, but he incorporated Wilkins's standards of discourse into his own *Advice to Candidates for the Ministry* (see pp. 48–51 and 61 above).

Morton represents the significant link between the Society and Defoe. For it was at the Dissenters' Academy that Defoe was introduced to the methods and aims of Baconian science; here he first heard of the influence of experimental thinking upon style. At an impressionable age, Defoe was taught that 'the Accuracy of Speech be not more minded than the Efficacy' and to keep discourses 'mostly Practical, both as to the Subjects, and Manner of Handling'. At Morton's Academy Defoe learnt to dispose of 'Things prudently (not Words curiously)'. '*Things* and not *Words*', Morton had insisted, quoting Cato the Elder's phrase as it had been quoted by Bacon and to the same purpose. It is significant that when Defoe declares that the 'Knowledge of things, not words, make a schollar', he does so in a work replete with references to Morton's practical attitude to learning (*Gentleman*, p. 212). Long before Defoe became a merchant and writer of economic tracts he was taught the value of plain, precise expression. Acknowledging this early influence upon his habit of mind and mode of communication, Defoe frequently praises his former teacher for instilling in him the love for '*plain Things in a plain Form*' (*Review* VIII: 199; for Defoe's enduring devotion to Morton see pp. 38–9 and 55–7 above). There can be little doubt that the main impulse to Defoe's later conviction that 'nothing can be ... more useful to the publick services than plain, naked, and unbyasst accounts both of persons and things' came from the Baconian philosophy of things as it had been taught by Morton (Healey 1969: 256).

In *Crusoe* the primacy of observation and experience controls the style. Crusoe studies nature and then describes his experience in plain, ordinary words. First priority is given to truthfulness; 'the subject', as Crusoe confirms in volume three, is 'honesty'. Frequently he not so much describes as merely names or lists things. Fidelity to fact dictates that he tabulates every single item salvaged from the shipwreck. He gives us an accurate inventory of his storeroom in the same manner in which he registers and adds up the total of the savages killed. Crusoe specifies time and space, he numbers, weighs and measures in order to take stock of every minute detail of his island as 'it is in fact'.

Crusoe consciously strives to give an objective, accurate and reliable account. Describing the sowing of corn, he records:

> Finding my first Seed did not grow, which I easily imagin'd was by the Drought, I fought [*sic*] for a moister Piece of Ground to make another Trial in, and I dug up a Piece of Ground near my new Bower, and sow'd the rest of my Seed in February, a little before the *Vernal Equinox*; and this having the rainy Months of *March* and *April* to water it, sprung up very pleasantly, and yielded a very good Crop.
>
> (p. 105)

The only thing to be remarked about that typical passage is that it is really unremarkable and contains the facts that we would expect. In his regard for truth Crusoe gives an exact report, with a mass of supporting details. The passage just quoted corresponds to the reports that the members of the Society regularly received from the 'artificers'. Here to compare is 'a written account of potatoes from [Mr Boyle's] gardiner, which was ordered to be entered' in the official records of the Society on 8 April 1663:

> I have, according to your desire, sent a box of potatoes. My care hath been to make choice of such, that are fit to set without cutting ... If you are minded to have great store of small roots, which are fittest to set, you may cause them to lay down the branches in the month before named, and cover them with earth three or four inches thick ... Now the season for digging the ground is in April or May, but I hold it best the latter end of April; and when they dig the ground, let them pick out as many as they can find, small and great, and yet there will be enough for the next crop left.
>
> (Birch 1756, I: 216–17)

The chief purpose of both accounts is to impart useful information. Both reports, or 'histories', were collected 'by the plainest Method, and from the plainest Information ... from ... *experienc'd* [that is, practical] *Men* of the most unaffect'd, and most unartificial kinds of life' (Sprat 1959: 257). It is in his role of scientific observer that Crusoe informs us that he found a 'moister Piece of Ground', that he made 'another Trial', and succeeded 'a little before the *Vernal Equinox*'; his plain style seeks to reflect the accuracy and reliability of his experiments.

There is one aspect of Crusoe which links him with the whole programme of the Baconian philosophy of things. The first half of the book is almost exclusively concerned with Crusoe's making and improvement of his world. Together with the experimental scientists, Crusoe is preoccupied with 'the world of things' and, as in their case, it is not just 'things' but more specifically 'the *making* of things' that demands his attention. Crusoe's concern with 'making' deeply affects the style of his narrative. In his account of his 'Mastery of every mechanick Art' (which in my edition falls between pages 47 and 153) the verb 'to make' is used 162 times. In places the word is used over and over, as on pages 73–4, 117, 120, with the biggest cluster of all when Crusoe tells us how he made his clothes and his umbrella (pp. 134–5). The word is used 17 times in this brief episode. The idea of making and the verb 'to make' dominate Crusoe's solitary years on the island. With the discovery of the footprint, the focus shifts from making to the protection of things, and so the prominence of the word recedes. As other students of *Crusoe* have pointed out, this is a real turning-point in several ways; my analysis (and breakdown of the text) confirms these observations.

Admittedly, Defoe does not often have, or take, the time to select his words imaginatively or creatively to achieve literary grace. He is guilty of hastiness and even carelessness. But it is also true that once he finds the perfect expression that fits the thought, he sees no reason why he should not repeat it. And for Crusoe, the maker of things, there is no word that would better match the action than 'to make'. The same could be said of the repeated use of 'thing'. In those representative pages we encounter it 142 times. Whether we agree that these repetitions are intentional or not, the effect is certainly one of a plain, unadorned, 'sachlich' or thing-like prose.

Curiously enough, Crusoe seldom leaves his exact measurements unqualified. So, for example, the already quoted passage where Crusoe describes the pitching of his tent, reads: 'This Plain was not above an Hundred Yards broad, and about twice as long, and lay like a Green before my Door' (pp. 58–9). Crusoe's account abounds in '*Approximating and alternative* counts'.[18] Defoe follows the Society's advice for scrupulous regard for truth but then withdraws, as if too much 'Mathematical plainness' in Crusoe's story would be unlikely. We can, of course, only speculate, and the most likely explanation is that for the sake of credibility Defoe added *post factum* approximating terms that blur the specificity. With one hand he gives accuracy, while with the other he disturbs the clarity to render a more life-like reality.

Believing that the plain facts should speak for themselves, the Baconian scientists avoided the use of figurative language. Crusoe, or his creator, shared this belief. On the rare occasions when metaphor and imagery are used, they are employed to a limited purpose. Crusoe uses pictorial language to dramatise an incident: after the discovery of the footprint he tells us that he 'fled into [his cave] like one pursued ... for never frighted Hare fled to Cover, or Fox to Earth, with more Terror of Mind than I to this Retreat' (p. 154). Sometimes metaphor is used to express a psychological reaction. So, when Crusoe finally succeeds in making some indifferent-looking pots, he reports that his primitive method was identical to that of 'the Children [who] make Dirt-Pies, or as a Woman would make Pies, that never learn'd to raise Past [that is, make pastry]' (p. 121). The comparison conveys admirably both the determination and joy, as well as the clumsiness of the untaught, unprofessional experimenter. Having thoughtlessly constructed a far too heavy boat, Crusoe reprimands himself. He is both angry and disappointed with himself for having acted so illogically and entirely without 'fore-thought', in fact, 'the most like a Fool' (p. 126).

Most frequently figurative language serves to express a moral or an emotional idea. Crusoe rejects money as 'a Drug' that fills man's heads with vaporous ambitions. On the island he judges gold worth no more than 'the Dirt under my Feet' (pp. 57 and 193) – once he returns to civilisation and the possibility of exchange, Crusoe quickly readjusts

---

[18] Rogers 1979: 122–3.

this point of view. He refers to himself as 'a Prisoner lock'd up with the Eternal Bars and Bolts of the Ocean' (p. 113); more often though he sees himself as 'the Lord and Governour' over his physical surroundings. Are these metaphors? I am reminded of Petty who could describe an anatomy theatre as '(without metaphor) a temple of God' (in Hill 1965: 92). Similarly, Crusoe's island is 'without metaphor' his world.

Not surprisingly, we find God's omnipotence expressed in metaphor. God is 'the wise Governour of all Things' (p. 197), or 'the great Maker of all Things' (p. 216); in contrast we, utterly depending on his benign will, are as 'the Clay in the Hand of the Potter' (p. 210). Crusoe in his efforts at pottery, at making things and governing his complete but microcosmic world, is shown to emulate the perfect workings of the 'divine artificer'. Crusoe's real but human, and hence imperfect attempts reflect God's complete order, comprehensible to him only in metaphors and similitude.

When Crusoe uses metaphor and imagery it is not to transform or sublimate his experience: rather it is to root it more firmly in reality. We know of Defoe's skill in summing up ideas in pithy phrases or aphorisms. This special gift he handed down to Crusoe, who uses imagery to turn out such succinct observations as 'Peace and Plenty [are] the Hand-maids of a middle Fortune'; 'in easy Circumstances' we will be 'sliding gently thro' the World, and sensibly tasting the Sweets of living' (p. 5); 'Life is a Chequer-Work of Providence' (p. 156); 'the Evil which in it self we seek most to shun ... is oftentimes the very Means or Door of our Deliverance' (p. 181). Above all, these proverbial phrases are symptomatic of Defoe's predilection to teach and dispel misconceived ideas. Expressive of Defoe's moral outlook, these wise, terse statements sound like proverbs; they are charged with practical experience derived from everyday life. In the way Defoe uses, adapts and plays with the proverb 'necessity is the mother of invention', he demonstrates his first-hand acquaintance with the Royal Society's precept that 'a man's own experience is the best part of his Learning' (p. 106 above).

Throughout, Crusoe aims to give us a concise, step-by-step account of his experience. But paradoxically his very attempt at honest witnessing creates its own peculiar complexity. He aims at transcribing truthfully the great difficulties he encountered in baking bread, constructing the boat, making the umbrella, etc. In his effort to convey as honestly and directly as possible the physical and mental strain, he is forced to repeat himself; he digresses, errs and stumbles stylistically as he had stumbled literally. The style re-enacts the experience. Take for example the moment when Crusoe 'studies' how to make a pot that would hold liquid:

> It happen'd after some time, making a pretty large Fire for cooking my Meat, when I went to put it out after I had done with it, I found a broken Piece of one of my Earthen-ware Vessels in the Fire, burnt as hard as a Stone, and red as a Tile. I was agreeably surpris'd to see it, and said to my self, that certainly they might be made to burn whole if they would burn broken.
> ... I had no Notion of a Kiln, such as the Potters burn in, or of glazing them with Lead, tho' I had some Lead to do it with; but I plac'd three large Pipkins, and two or three Pots in a Pile one upon another, and plac'd my Fire-wood all round it with a great Heap of Embers under them, I ply'd the Fire with fresh Fuel round the outside, and upon the top, till I saw the Pots in the inside red hot quite thro', and observ'd that they did not crack at all; when I saw them clear red, I let them stand in that Heat about 5 or 6 Hours, till I found one of them, tho' it did not crack, did melt or run, for the Sand which was mixed with the Clay melted by the violence of the Heat, and would have run into Glass if I had gone on, so I slack'd my Fire gradually till the Pots began to abate of the red Colour, and watching them all Night, that I might not let the Fire abate too fast, in the Morning I had three very good, I will not say handsome Pipkins ...
>
> (pp. 120–1)

Truthfulness to fact and the process of discovery demands asides, indecisions, repetitions. Details and ideas are strung together as and when they occur. The subordinate clauses of the last sentence are piled one upon another like the pots in the fire. It is as if honesty prevented any more strict or graceful ordering of words. But then, it is exactly in these stylistic imperfections that we find the perfect rendering of the moment.[19] The account of the laborious and lengthy experiment tries our patience, the way it taxed Crusoe's. Style and structure let us not only witness but participate in the method of trial and error. This analysis of a particular paragraph serves to show in detail the impress

---

[19] Rogers, though arguing from a different aspect, arrived at a similar conclusion, 1979: 124.

of the Society's demand for a direct rendering of experience on Defoe's writing.

Looking back to the seventeenth century, Defoe firmly believes that the 'Knowledge of things, not words, make a schollar' (*Gentleman*, p. 212). In *Crusoe* words and works, words and things are matched with such astonishing directness that not perspicuity but 'the anarchy of thought and the chaos of the mind' are preserved.[20] Rather than giving an order of words, Defoe recaptures the *process of ordering* a higgledy-piggledy world of impressions. He applies the pair of words in an entirely new way. Morton's (or, to be precise, Bacon's) advice on style that 'it take hold of things' has found a novel interpretation. Equating plainness with sincerity, art with deception, Defoe dismisses artifice and acts 'the honest rather than the artfull part' (Healey 1969: 256). While his 'pointed Truth ... and downright Plainness' were not always a success, in *Crusoe* textual and stylistic truthfulness to fact embody the very essence of the narrative's enduring attraction. Here the experimental scientists' pursuit of truth evolved into a profoundly imaginative and creative activity. Rejecting the 'artfull', Defoe, in James Joyce's words, 'devised for himself an artistic form which is perhaps without precedent' (Joyce 1964: 3–27). The 'natural infirmity of homely plain writing' could not command words, but it conveys convincingly the conquest over things. Plain fact and plain style mysteriously cohere to render the 'brazen' world in the very act of being created.

[20] From Dryden's 'The State of Innocence':
> From words and things, ill sorted and misjoined;
> The anarchy of thought and chaos of the mind.
>
> (Dryden 1883, V: 147)

7

A New Voyage Round the World: Defoe the traveller-scientist by sea

The traveller and the Royal Society

Although *Crusoe* 1 is Defoe's best-loved travel-adventure story, very little travelling takes place in this book. For actual accounts of travel we have to turn to *Crusoe* 2, *Captain Singleton*, *A New Voyage Round the World* and *A Tour thro' the Whole Island of Great Britain*. The last two of these works will be the subject of the final chapters of this book.[1] Published in the mid-1720s, these travel accounts appeared at the end of the author's most productive period when he was at the height of his career as creative writer. My argument will be that in these two travel books, one fictional, the other factual, we have Defoe's skilled application of the experimental scientists' concern with 'useful knowledge for the good of all'. The following brief outline of the Royal Society's influence upon the voyage literature of the Restoration period and early eighteenth century will serve as introduction to the discussion of *A New Voyage* and the *Tour*.

In the Royal Society's pursuit of knowledge, the traveller played an important role from the start. Robert Boyle, John Ray, John Aubrey, Edward Lhwyd, John Woodward, Robert Plot, John Evelyn and Ralph Thoresby were among those who, believing in the need for research

---

[1] *Crusoe* 2 and *Captain Singleton* are the subject of a forthcoming article.

based on direct contact with nature, journeyed through England and various parts of Europe to report faithfully on the natural and man-made world. The Fellows of the Royal Society felt, however, that it was in the first instance not the scholar but the practical, 'unlettered' traveller with his frequent journeys to all corners of the world who was in the best position to supply them with an abundance of personal observations. Shortly after its foundation the Society requested therefore 'Master Rooke ... to think upon and set down some *Directions* for *Sea-men* going into the *East & West-Indies*'. Lawrence Rooke (1622–62) compiled his *Catalogue of Directions* for travellers by sea for 'the better to capacitate them for making such observations abroad, as may be pertinent and suitable for [the Society's] purpose'.[2] With these *Directions* the Fellows instructed sea-bound voyagers in the guiding principles of their philosophy: they ought 'to study *Nature* rather than *Books*, and to compose such a History of Her, as may hereafter serve to build a Solid and Useful Philosophy upon'. These are all Baconian motifs. Right from its inception, then, the Royal Society appealed to 'Seamen, Travellers, Tradesmen, and Merchants' to contribute with their personal experience 'to the making of a Natural History in general' (Sprat 1959: 155–6).

The Society's *Catalogue of Directions*, which was published in the *Philosophical Transactions* for January 1665–6, instructed the sea-bound voyager in such details as 'to observe the Declination of the *Compass*'; 'to carry *Dipping Needles*'; 'to remark carefully the Ebbings and Flowings of the Sea'; 'to make Plots and Draughts of prospects of Coasts'; 'to keep a Register of all changes of Wind and Weather at all houres'. Within the same year Robert Boyle extended these instructions with similar advice for travellers by land. Boyle's advice, which bears the title 'General Heads for a Natural History of a Country, Great or Small; Drawn out for the Use of Travellers and Navigators' emphasises the importance of moving from physical investigations to the study of the customs, education, and trade of the inhabitants.[3]

---

[2] *Philosophical Transactions* 1665–6, I: 141. For the following account of the relationship between the experimental scientist and the Restoration traveller I am indebted to Frantz 1968, C. Webster 1975: 420–80, Shapiro 1983: 124ff and Edwards 1994. For an excerpt from the *Directions* see pp. 106–7 above.

[3] Boyle's original outline appeared in the *Philosophical Transactions* 1665–6, I: 186–9. In 1692, shortly before his death, 'The General Heads for the Natural History of a Country' were

We are not short of proof that the traveller of the period followed the Society's instructions to observe and record physical phenomena. Responding to the Fellows' call for assistance to increase their stock of knowledge, voyagers like William Dampier, John Narborough, Woodes Rogers, Lionel Wafer, A. O. Exquemelin, Amédée Frézier, equipped themselves with the Fellows' *Directions* and set out to explore the world. From their accounts it is evident that the system of thought advocated by the Baconian scientists profoundly influenced their method of gathering information. The reports of these voyagers abound in statements assuring the reader that the writer was 'guided wholly by *Matter of Fact*'. Thus, Lionel Wafer in his account of his journey across the Isthmus of Darien stressed that he 'was there upon the spot' and that his report is 'matter of Fact ... related without the Embellishment of [his] owne Brain' (Wafer 1934, Appendix I: 150–1). Similarly, Woodes Rogers in the Introduction to *A Cruising Voyage Round the World* explicitly states that he knew that it is 'generally expected, that when far distant Voyages are printed, they should contain new and wonderful Discoveries, with surprizing Accounts of People and Animals' but that he will avoid such reports since 'it is not reasonable to expect such Accounts here' (Woodes Rogers 1718: xiii).[4] The lesson which the Society had taught and which the traveller of this period had taken to heart was that nothing was recorded 'but what hath due warrant from Observations; and those both carefully made, and faithfully related' (Woodward 1695: 2).

Certainly, the Restoration traveller was not the first meticulously to record and collect; what was different and characteristic of the period from 1660 to the early eighteenth century was that the traveller could

published with additions 'by another Hand'.

The importance of travel in the Society's way of thinking may also be gauged from the space devoted in the *Philosophical Transactions* – particularly between 1700 and 1720 – to accounts of travel. The abridged edition of the *Transactions* from 1700 to 1720 devotes a long section (vol. V, chap. 3) to 'Travels and Voyages'. *A Collection of Curious Travels, Voyages, Antiquities, and Natural Histories of Countries* constitutes volume two of the three-volume edition of *Miscellanea Curiosa. Containing a Collection of Some of the Principal Phenomena in Nature, Accounted for by the Greatest Philosophers of this Age; Being the Most Valuable Discourses, Read and Delivered to the Royal Society, for the Advancement of Physical and Mathematical Knowledge* (London, 1705–7).

[4] Characteristic of the traveller-scientist, Rogers continued : 'The rest is from my own knowledge, being a Description of those Places we were at, with such Remarks as occur'd to my Observations, and that I thought might be useful to them who may hereafter trade to those parts' (Rogers 1718: xiv).

for the first time engage with research scientists in an 'universal correspondence', and that they were both working towards the advancement of useful knowledge for the benefit of man. Dedicating his *New Voyage Round the World* (London, 1697) to the President of the Royal Society, Dampier commented:

> I [cannot] think this plain Piece of mine, deserves a place among your more Curious Collections ... Yet dare I avow, according to my narrow Sphere and poor Abilities, *a hearty Zeal for the promoting of useful Knowledge, and of any thing that may never so remotely tend to my Countries Advantage*: And I must own an Ambition of transmitting to the Publick through your Hands, these Essays I have made toward those great Ends, of which you are so deservedly esteemed the Patron. This hath been my Design in this Publication, *being desirous to bring in my Gleanings from here and there in Remote Regions, to that general Magazine, of the Knowledge of Foreign Parts*, which the Royal Society thought you most worthy the Custody of, when they chose you for their President ...
>
> (Dampier 1906, I: 17–18; my italics)

Captain William Dampier (1652–1715), without doubt the most celebrated traveller and buccaneer of this period, had recorded his circumnavigation of the globe in his hugely successful *A New Voyage Round the World* (a title which Defoe later borrowed). By nature endowed with a penetrating curiosity, a penchant for detail and a talent for describing things exactly as he perceived them, Dampier rapidly became the Restoration traveller *par excellence*. Representative of his conscious effort for brief, mathematical precision is the following description of a shark:

> Among [the sharks] we caught one which was 11 Foot long. The space between the two Eyes was 20 Inches, and 18 Inches from one Corner of his Mouth to the other. Its Maw was like a Leather Sack, very thick, and so tough that a sharp Knife could scarce cut it ...
>
> (Dampier 1906, II: 427)

As his carefully kept journals show, Dampier numbered, weighed, measured and whenever possible, also smelled and tasted physical reality. Thus, his description of anteaters ends typically with the observation that 'they smell very strong of Ants, and taste much stronger, for I have eaten of them'. With similar conciseness he reported on, and often drew, plants, insects, fish, indeed everything that he could lay hold of.

Dampier's declared zeal for the promotion of useful knowledge is characteristic of the Baconian Restoration traveller, as is his ambitious aim of contributing with his 'Gleanings from here and there' to the Society's 'general Magazine of the Knowledge of Foreign Parts'. Following the Society's instructions, Dampier and other voyagers compiled histories of the soil, the air, the atmosphere, the weather (Dampier's history of trade-winds became, as J. K. Laughton, author of the *DNB* article on Dampier points out, 'one of the most valuable of all the "pre-scientific" essays on meteorological geography, and is even now deserving of close study' (XIV: 4); for Edmond Halley's history of trade-winds see *Transactions* 1686–7, XVI: 153–68).[5] They reported on comets, earthquakes, the ebb and flow of the sea, on fossilised snails and fish; they brought in animals dead and sometimes alive.[6] Responding to the Society's plea for help, the traveller of this period rapidly developed into an amateur scientist whose faithful reports were considered by the virtuosi as 'highly conducive' to the augmenting of their knowledge and the welfare of mankind. When the Fellows needed precise information, they turned to the traveller-scientist. Ray, Hooke, Lhwyd, and Woodward acknowledged with gratitude observations they had received from the 'merchant voyagers'. So, for example, Ray wrote that 'an Experiment ... occurred to [him], which much confirmed [him] in the belief and perswasion of the Truth of those Histories and Relations which Writers and Travellers have delivered to us concerning dropping Trees in *Ferro, S. Thome, Guiny* etc. of which before I was somewhat diffident' (Ray 1693: 113; and see Woodward 1695: Preface and p. 5). Recognising the great value of the travellers' eye-witness accounts, the Fellows frequently printed these in their *Transactions*, where they appear next to the scholarly investigations of a Wren or a Boyle.[7] The travellers working in the service of the Royal Society evidently understood that their aim was 'not to direct or gratify

---

[5] For Edmond Halley's important contribution to the history of trade-winds see Thrower 1978.

[6] See the entry in the *Philosophical Transactions* for April 1698, 'Carigueya, *seu* Marsupiale Americanum, or, *The* Anatomy *of an* Opossum', where we read: 'This *Animal* was brought from *Virginia*, and presented to the *Royal Society*, by *Will Bird*, Esq.; and kept alive in their Repository for some time', *Philosophical Transactions* 1698, XX: 105–64.

[7] *Philosophical Transactions* 1668–9, III: 717–22, 722–5, 792–5, 817–25; 1693, XVII: 781–95, 941–8, 978–99; 1694, XVIII: 121–35; 1695–7, XIX: 83–110, 129–60, 225–8; 1698, XX: 167–8; 1699, XXI: 436–42; 1700–1, XXII: 536–43, 729–38; 1706–7, XXV: 2,423–34.

the Curious Reader' (Dampier) but simply to instruct. To this end they employed the style recommended by the Fellows that reflected '*the severe, full and punctual Truth*' of their inquiries. It is interesting that the voyagers not only adopted the Society's stylistic principles but found it necessary to draw the reader's attention to this fact. Reading their accounts, again and again we come across statements confirming their deliberate preference for plainness in both content and form of their writing. Defending his unadorned method of recording his observations, Woodes Rogers asserted that he had 'had no time to polish the Stile', while Dampier declared that his main purpose was rather to give 'a Plain and Just Account of the true Nature and State of the Things Described, than of a Polite and Rhetorical Narrative' (Rogers 1718: xiv, Dampier 1906, II: 342).

One final important point needs to be made. From the outset, the Fellows encouraged voyagers to extend their investigations of natural phenomena to economic observations. The link between science, travel, trade and colonisation was thoroughly consistent with the Baconian injunction to use the histories of nature 'to the enlarging of the Mind and Empire of Man'. Moreover, the intersection between scientific exploration and exploitation of natural resources was a singularly attractive one, holding, as it did, the possibilities of immense national power and wealth. When the *Catalogue of Directions* was reprinted in Churchill's *Collection of Voyages and Travels* (1704), the writer of the Introductory Discourse once again highlights 'the Advantages the Publick receives by Navigation, and the faithful Journals and Accounts of Travellers'. The intersection between science, travel and useful information is 'natural', since 'no Man can read the one without being sensible of the other'. The writer goes on:

> Trade is rais'd to the highest pitch, each Part of the World supplying the other with what it wants, and bringing home what is accounted most precious and valuable ... *To conclude, the Empire of Europe is now extended to the utmost Bounds of the Earth, where several of its Nations have Conquests and Colonies.* These and many more are the Advantages drawn from the Labours of those, who expose themselves to the Dangers of the vast Ocean, and of unknown Nations, which those who sit still at home abundantly reap in every kind : and the Relation of one Traveller is an Incentive to stir up another to imitate him, whilst the rest of Mankind, in their accounts without stirring a foot,

compass the Earth and Seas, visit all Countries, and converse with all Nations.

(Churchill 1704, I: lxxiii, my italics)

If this passage succinctly describes the close relationship that there existed between science, travel and trade, it also calls to mind one of Defoe's best-known sayings concerning the acquisition of practical knowledge. Advising those who have not had the opportunity to travel in their youth, Defoe recommends that they 'make the tour of the world in books ... go round the globe with Dampier and Rogers ... [and so] make all distant places near to [them] in [their] reviewing the voiages of those that saw them' (*Gentleman*, p. 225). That Defoe followed his own advice and not only supplemented his experience with a 'tour of the world in books' but specifically with books written in accordance with the Society's guidelines to travellers will become clear in the remaining pages of this study.

Defoe the traveller-scientist by sea

The narrative of Defoe's *A New Voyage Round the World* ('1725', for 1724) falls neatly into two equal parts : the first half of the book treats a merchant voyage to the Philippines and Pacific Ocean, while the second part mainly focuses on South America.[8] Significant for the present discussion of the influences upon *A New Voyage* is that Defoe applies in each section a distinctly different method of using his source material. In the first part of the book the author concentrates on the benefits derived from going round the world by 'a course never taken before'. Defoe's argument here is that by sailing first to the Orient, trading ships could sell their English merchandise in the Philippines 'double the advantage we had already made' (p. 269); the ships could then re-load with a rich cargo of silver, pearls and chinaware which they could again sell at a huge profit to the Spanish colonists. The main purpose of this part of the narrative is to demonstrate that a trans-Pacific trade was not only possible but very profitable if English traders imitated Defoe's example and took the journey in two stages. To make his case Defoe needs to give not the account of the voyage

---

[8] For the following discussion of *A New Voyage* I have used the Bohn's British Classics edition of 1856. All references will be to this edition and will be given in the text in parentheses.

but rather the account of arriving, of selling and re-stocking, and, finally, of clearing a huge double profit. Throughout this part of the book the comments of the narrator (a thin disguise for Defoe himself) on the physiography of the ports of call, the animals and plants are general rather than specific. The same is true of the inhabitants who are invariably described as either peace-loving or treacherous. Although Defoe pretends to remove misconceptions of the inhabitants of Madagascar (the false 'ideas of the figure those people made in Madagascar, from the common report in England'), in fact he does not go into much detail and merely observes that 'the people [are] wild, naked, black, barbarous, perfectly untractable, and insensible of any state of life better than their own' (p. 256).[9] The information given in this part of the story is the most elementary; it could have been found in countless books on travel of any period.

Although at the time when *A New Voyage* was published in 1724 Defoe had a great number of reports on East India voyages at his disposal, no single specific source could so far be discovered for the voyage to the Philippines. We derive our knowledge of Defoe's extensive familiarity with the literature of voyages to the Orient from *Robinson Crusoe* and *Captain Singleton*, not from *A New Voyage*. When Defoe came to write this part of the book he most likely worked from memory, re-using research he had carried out for his earlier travel books. I would further like to suggest that since his argument for the benefits that derived from an easterly circumnavigation did not depend on precise information regarding the inhabitants or the animal and plant life, Defoe decided to give no more than a loose imitation of a traveller's report.

In contrast to the first section of the voyage, when describing that across the Pacific Ocean Defoe was faced with an irritating lack of information. At the beginning of the eighteenth century English oceanic history was still in its infancy and on Edmond Halley's superb world map of 1700 the largest blank space was the Pacific (see fig. 3).[10] Confronted with this *carte blanche* Defoe was forced to fabricate most of his material, a challenge he naturally fulfilled to suit his own end. Thus,

---

[9] For a more detailed discussion of Defoe's attitude to the savage see pp. 120–1 above.
[10] For an insightful discussion of Halley's contribution see 'The achievement of the English voyages, 1650–1800' by Glyndwr Williams in Howse 1990.

3 Edmond Halley's *Magnetic Chart* (1700).

the merchant captain 'discovers' many islands rich in natural resources, particularly gold and pearls. Everywhere the sailors go they meet 'tractable and courteous' inhabitants, who could 'be made easily subservient and assistant to any European nation that would come to make settlements among them' (p. 330). It is interesting that whenever possible Defoe weaves into his fictional text such scientifically established information as he can lay hold of. Since A. W. Secord's careful spadework more than sixty years ago, we are aware 'that Defoe's *New Voyage* is, in those parts dealing with the Moluccas and Australia, largely an ill-disguised imitation of Dampier's report of his Australian voyage' (Secord 1924: 153). However, indisputable as Defoe's use of Dampier is, it is very difficult to pin-point the extent of his debt, and Secord, and later Bonner, could trace only processed borrowings. In part one of *A New Voyage* Defoe's dependence on his sources is pervasive rather than direct. Instances where we are permitted to glimpse the origin of his knowledge are rare. Here is one of them: Defoe lets his captain-narrator observe that being 'then in the latitude of 67 degrees south, which I suppose is the farthest southern latitude that any European ship ever saw in those seas' (p. 323). This speculation is of course perfectly correct since it is based on Woodes Rogers's account of *A Cruising Voyage Round the World* which had taken him to 'Lat. 61.53. Long. W. from Lond. 79.58. being the furthest we ran this way, and for ought we know the furthest that any one has yet been to the Southward' (Rogers 1718: 108–9).

The most important point, or, as Defoe puts it 'the chief design of the whole voyage' is to be found in part two of the book. It has long been accepted that the South American trading scheme is a fictional reworking of ideas Defoe had entertained for at least twenty years. Several issues of the *Review* were devoted to the subject of colonisation and commercial exploitation of South America. In a letter to Robert Harley, dated 23 July 1711, Defoe outlined in greatest detail his plan for an English trading colony in Chile linked by routes across the Andes to a sister colony in Patagonia (Argentina) at Camerones Bay (Healey 1969: 349). Although the government took no action, Defoe's enthusiasm was not dampened and he renewed his proposal in his *Historical Account of the Voyages and Adventures of Sir Walter Raleigh* (1720), *History of Arts and Sciences* (1725), and once again, in one of the last works to be published, the *Plan of the English Commerce* (1728). In *A New Voyage* Defoe created for himself the opportunity of making his long-cherished dream come true in the reality of his fiction. There can be little doubt that for the ideas themselves Defoe was his own source.

The author's declared purpose of *A New Voyage* is

> to recommend that part of America as the best and most advantageous part of the whole globe for an English colony, the climate, the soil, and, above all, the easy communication with the mountains of Chili, recommending it beyond any place that I ever saw or read of...
>
> (p. 458)

In order to carry out his mission of presenting to the English nation a realisable, profitable trans-Andean trading scheme, Defoe seeks verification in works generally accepted to be reliable. The most important of the established sources to date is Moll's map of 'the Kingdom of Chili' in his *The Compleat Geographer* (1709).[11] Hermann Moll, a Dutchman, was one of the most important London cartographers of Defoe's day. Defoe had already made use of Moll's work, his world map printed in the front of the first volume of Dampier's *Voyages*, in the composition of *Crusoe* and *Singleton*; a few years later, he would return to Moll for details for the *Tour*. As far as *A New Voyage* is concerned, Moll's map furnished Defoe with such information as made him conclude that his life-long dream could be turned into a reality. Using Moll, he argued confidently that there existed overland passages from Chile to Peru through the Andes; that these passages permitted carriages; and, finally, that there were large navigable rivers flowing from the Andes into Camerones Bay. Today we know that Moll's map was wrong and that Defoe was in fact misled into thinking that there was a navigable Rio Camerones; he also underestimated the hazards and distances of the trans-Andean trek. This does not diminish the significant point that Defoe sincerely believed that in Moll's map he had found scientific evidence of the realisability of his trans-Andean trading scheme.

As mentioned above, one of the books in Defoe's mind when composing *A New Voyage* was Woodes Rogers's *A Cruising Round the World*.[12] Not only does Woodes Rogers give an extensive 'Account of

---

[11] For Defoe's indebtedness to Moll see Fishman 1973.
[12] Defoe's use of Woodes Rogers in *Robinson Crusoe* is well documented in Pat Rogers 1979.

the River La Plata' but he also supplied useful information on an existing route from Buenos Aires to Chile. Furthermore, in a section 'Chili describ'd', Rogers gives a detailed summary of Alonso de Ovalle's 'An Historical Relation of the Kingdom of Chile' which had appeared in the *Collection of Voyages* by the Churchill brothers. At the beginning of the eighteenth century Ovalle was one of the most respected sources of information on South America, his 'Natural History of Chile' being considered 'a Model for most Relations of that kind' (Churchill 1704, III). It is hard to believe that Defoe, curious as he was about South America, would not have been familiar with the full version of Ovalle's 'Historical Relation'.

Other accepted direct sources for the South American adventure are A. O. Exquemelin's *The Bucaniers of America* (1678; English translation 1684–5), and Lionel Wafer's *A New Voyage and Description of the Isthmus of America* (1699). Defoe, who possessed copies of these books, appears to have used them for his description of the overland crossing of South America. His method of borrowing is, however, concealed: 'while he leans fairly heavily on Wafer and Exquemelin, in this part of the book, he is not dependent on them and remains firmly in charge of his narrative'.[13]

To this list of source materials I would now like to offer one other travel account. Although the influence of John Narborough on *A New Voyage* has been noted before, the full extent and deeper significance of Defoe's indebtedness seems to have gone undetected so far.[14] Admiral John Narborough had been commissioned in 1669 by Charles II (the then patron of the Royal Society) to carry out an exploration expedition to the South Seas. He kept a detailed journal which was later published in *An Account of several late Voyages and Discoveries to the South and North towards the Streights of Magellan* ... (1694). Reviewed in the *Philosophical Transactions*, the *Account* was highly praised for its commitment to the ideals of the Society, that is, that it was written 'in order to improve and compleat Geography from Original Authentick Records, and Memoirs of Eye-Witnesses'. Sir John Narborough's

---

[13] Jack 1961: 335. Defoe owned the 1699 edition of the *Bucaniers* containing accounts by Exquemelin and Basil Ringrose.

[14] Secord 1924: 153, Fishman 1973: 231. Peter Earle wrote in 1976 that 'Narborough's account was one of Defoe's main sources for South America' (1976: n. 58 p. 298).

faithful observations, writes the reviewer, 'must needs contain many uncommon and useful Things upon most of the Heads of Natural and Mathematical Sciences, as well as Trade and other Profitable Knowledge, which contribute to the enlarging of the Mind and Empire of Man'. The passage that would have particularly interested Defoe runs:

> Sir *John Narborough* ... delineates and describes the Coasts of *Patagonia* and *Chili*, together with the Streights of *Magellan*, of which he took most exact Draughts, going frequently on Shoar, and up into the Country, observing the Products of the Land, the Manners and Tempers of the *People*, in order to promote and settle an Advantagious *Trade*, especially in the Golden Country of *Chili*, esteemed more Rich than *Peru* itself.
> (*Philosophical Transactions* XVII: 166–7)

Further proof of Narborough's endorsement of the Society's *Catalogue of Directions* is to be found in his own instructions to Captain Fleming who accompanied him on the voyage. Here we read:

> You are hereby required ... to observe all Headlands, Islands, Bays, Havens, Roads, Mouths of Rivers, Rocks, Shoals, Soundings, Courses of Tides, flowings and settings of Currents ... and also you are to take notice of all Trade-Winds, etc. you meet with, and of the Weather ... You are in all places where you land to observe the nature of the Soil, and what Fruits, Woods, Grain, Fowls, and Beasts it produces, and what Stones and Minerals, and what Fish the Rivers and the Sea doth abound with ... You are also to mark the temper and inclinations of the *Indian* Inhabitants, and where you can gain any Correspondence with them, you are to make them sensible of the great Power and Wealth of the Prince and Nation to whom you belong, and that you are sent on purpose to set on foot a Trade, and to make Friendship with them.
> (Narborough 1694: 10–11)

A clearer reflection of the Fellows' advice to travellers by sea could not be found.

Studying Narborough's and Defoe's texts side by side it becomes quickly evident that the traveller's personal observations profoundly influenced Defoe's fictional account. A brief comparison will speak for itself. Travelling along the east coast of South America, Narborough had gone on shore to observe the soil which he found

> *gravelly* and dry, in some Valleys well mixt with *black mould* . . . the Soil is *marly* and good, the Hills not very high, but plain large Downs, with Grass on them all over; digging in two or three places I found sandy dry ground near a foot deep, then Marle: In my opinion it might be made *excellent Corn-ground*, being ready to Till; 'tis very like the Land on *New-market* Heath . . .
>
> (pp. 26 and 37)

The corresponding passage in *A New Voyage* reads:

> As for the *soil*, that of the *hills* is *gravel*, and some stony; but that of the plains is a *black mould*, and in some places a rich loam, and some *marl*, all of which are tokens of fruitfulness . . . producing excellent corn.
>
> (p. 420)

Defoe depends for his information on Narborough. Obviously appreciating the accuracy of the firsthand report, he does not mind re-using Narborough's actual words (which for the sake of clarity I have italicised). Narborough had recorded that 'no woods could be seen', and Defoe faithfully echoes 'very little wood [is] to be seen anywhere'; again, Narborough had stumbled across an 'abundance of wild Pease', while Defoe records 'as for peas, they grow wild all over the country'. Defoe has no scruples in pilfering from Narborough's eyewitness account and applying his spoils to other regions of South America – justifying this method of borrowing he writes: 'for why may we not be allowed to suppose that the country on the same continent, and in the same latitude, should produce the same growth' (p. 328)? Curiously enough, Defoe feels compelled to reject Narborough's comparison of this fruitful part of the country with Newmarket Heath and instead compares it to Salisbury Plain. It appears that as far as the English countryside is concerned, he considered himself the expert and in his judgement it is Salisbury Plain which most resembles this idyllic place. After all, he had only recently personally studied both Newmarket and Salisbury and recorded the 'arcadian' beauty of the Plains in his *Tour*.[15]

The Society had directed that 'there must be a careful account given of the *Inhabitants* themselves, both *Natives* and *Strangers*' (Boyle); in Narborough's words just quoted above, the traveller must 'mark the temper and inclinations of the *Indian* Inhabitants'. Guided by these

---

[15] For Defoe's description of the 'arcadian' Salisbury Plain in the *Tour* see Letter 3 in Cole 1927.

instructions, Narborough had given a precise account of the inhabitants he had met. Observing the people on Elizabeth Island he had noted 'their clothing' which:

> is pieces of skins of Seals, and Guianacoes, and *Otter skins* sewed together, and *sewed soft*, their Garment is ... wrap[ped] about their Bodies, as a *Scottish Man doth his Plading*: they have a *Cap of the Skins of Fowls*, with the Feathers on; they have about their Feet pieces of Skins tied to keep their Feet from the Ground...
>
> (*Account*, p. 65, my italics)

Here to compare is Defoe's description of a Chilean. He

> was dressed in a jerkin made of *otter-skin*, like a doublet, a pair of long Spanish breeches ... very *soft* ... he had over it a mantle ... thrown about him *like a Scotsman's plaid*; he had shoes of a particular make, tied on like sandals ... He had on *a cap of the skin of some small beast* like a racoon, with a bit of the tail hanging out from the crown of his head backward...
>
> (p. 392, my italics)

The parallels between the two texts are unmistakable. Defoe's description of a Chilean comes with small changes from the pages of Narborough's journal. Thus, Defoe pretends to take on the role of those who with faithful and impartial diligence have observed and recorded the material world.

Is it coincidence that both Narborough and Defoe's captain should meet a hospitable Spaniard who takes them to his home where everything glitters with silver and gold?[16] We shall of course never know to what extent Defoe was inspired by this episode in Narborough's journal. Certain though is the similarity between the two reports and the following excerpt might as easily come from Narborough's as from Defoe's account:

> I had much Discourse with the *Spanish* Gentleman this day concerning *Baldavia* ... I asked him if there were any passage by Land from *Baldavia* to the other parts of *Chili*? [he] said there was, and [he] sent every Week, but they went with good Guards to go secure from the *Indians*.
>
> (*Account*, p. 94)

[16] The corresponding passages can be found in Narborough's *Account*, pp. 87–9 and in *A New Voyage*, pp. 371–2.

The important point is that both the actual and fictional traveller ask the same questions and that, with one exception, they receive the same answers. Both Narborough's and Defoe's captain learn from the Spaniard that there are passages that lead through the Andes to a fertile country stretching towards the Atlantic coast and that Chile is 'infinitely rich in gold'. Narborough's information which Defoe conveniently ignores is that 'the Natives do much hinder [Spaniards] getting of it; for they are at cruel Wars with them, and will not permit them to plant any thing near here about, nor at *Baldavia*, but they come and destroy it with Fire' (*Account*, p. 90).

In *A New Voyage* Defoe decides to hide his reliance on Narborough but in another work, his factual *History of Arts and Sciences*, published roughly at the same time as his fictional account, he honestly cites his source. Repeating once again his proposal for a settlement in South America, Defoe quotes directly from the *Account* and then acknowledges his debt to 'Sir *John Narborough*, a well known Person for his Experience in such Things as these' who gave his 'Testimony' and 'asserts it of his own Knowledge' (pp. 292 and 296). Finally, Defoe reveals his source and we know without a shred of a doubt that his propaganda piece for settlement in South America is based on Narborough's 'Testimony' and 'Experience in such Things as these'.[17]

In *A New Voyage* Defoe fuses fact and fiction and he lets his travellers come across Narborough's post erected in Port Desire in Patagonia:

> Here we found a post or cross, erected by Sir John Narborough, with a plate of copper nailed to it, and an inscription, signifying that he had taken possession of that country in the name of Charles the Second.
>
> Our men raised a shout for joy that they were in their own king's dominion, or as they said, in their own country; and indeed ... I never saw a country so much like England.
>
> (p. 421)

The passage underlying Defoe's fictional conquest is Narborough's entry into his journal for 25 March 1670:

> Gentlemen, You are by me desired to take notice, that this Day I

---

[17] Noteworthy, too, in this context is that the author's esteem of Narboroughs' firsthand testimony follows shortly upon Defoe's open declaration of allegiance to experimental science and his most extended direct borrowing from Robert Boyle (see pp. 76–7 above).

take possession of this Harbour and River of *Port Desire*, and of all the Land in this Country, on both Shores, for the use of his Majesty *King Charles the Second, of Great Britain*, and his Heirs; God save our King, and fired three Ordnance.

(*Account*, p. 40)

In *A New Voyage* Defoe uses the incident of Narborough formerly taking possession of Port Desire to underpin his own project, 'which was of no less consequence than of planting a new world, and settling new kingdoms, to the honour and advantage of my country' (p. 355). When he re-uses this episode in a *History of Arts and Science*, he phrases his argument like this: 'This I mention ... to intimate, that at least the English have as good a Title to [this country], as any other Nation whatsoever.' Evidently, Defoe shared the belief that the combined forces of science and trade could extend 'the Empire of *Europe* ... to the utmost Bounds of the Earth, where several of its Nations have Conquests and Colonies' (Churchill).

Projecting *A New Voyage* against Narborough's *Account*, we discover the extent of Defoe's indebtedness. Still more important is the fact that by imitating and copying from the reports of such respected travellers as Narborough, Dampier, Frézier, Wafer, Rogers, Exquemelin, Defoe reveals his knowledge of the virtuosi's pattern for travel. These traveller-scientists, as I have tried to show above, openly declared their allegiance to the Society's ideals of observing and recording natural phenomena. Anyone reading their accounts cannot avoid becoming familiar with the fundamental principles (that is, the Baconian principles), according to which they were written. Studying these travel writers, Defoe would have again and again come across the Fellows' demand for accurate and objective collection of data as set out in their *Catalogue of Directions* printed in the *Philosophical Transactions*. If Defoe did not have ready access to the *Transactions*, he could also have found the *Directions* in the popular Churchill *Collection of Voyages* (1704), where they had been reprinted in full. The aspect of this 'System of travel' which particularly appealed to Defoe was the recommendation to apply accurate data on the natural world to economic improvements and colonial expansion. That Defoe shared this belief is evident from such works as the *History of Trade* and *History of Arts and Sciences*; it is particularly prominent in the *Plan of the English Commerce* where Defoe's

49811"> where Defoe's 'Merchant-Adventurer' in pursuit of trade 'pushes on Discoveries, plants Colonies, and settles Commerce, even to all Parts of the World' (p. 26). Defoe's confidence in the powerful interaction between science, trade and colonisation is again demonstrated in part two of *A New Voyage*. It also informs his frequent exhortation to cultivate and plant nature. Defoe is convinced that South America is rich in mineral resources, particularly gold, but that exploration and exploitation had so far been held up by the ignorance and negligence of the Spaniards. 'All this vast treasure lies unregarded' by the Spaniards who do nothing to improve the land. His argument is that 'any nation in Europe that thinks fit to settle in it are free to do so' and he calls upon England to bestir itself and take possession of this part of the world (p. 402).[18]

What appears certain from the discussion above and from what will be made even more explicit in the following chapter on the *Tour* is that Defoe had an informed understanding of the Society's pattern for travel. He raids the auxiliary scientists' reliable evidence in order to strengthen his case, and his borrowings – sometimes explicit and sometimes oblique – permit us to trace the direction of his interests. One of the most decisive statements of allegiance to the New Sciences comes in the first pages of the book. The voyage is scarcely under way when the ship is blown off course and makes land on the east coast of South America. Finding himself on unexplored territory, the captain surveys the land and 'makes many observations' including '*An observation of the soil and climate of the continent of America, south of the river De la Plata; and how suitable to the genius, constitution, and the manner of living of Englishmen, and consequently for an English colony*' (p. 203). The captain's, or rather Defoe's, scientific History of the Soil and Climate is compiled in accordance with the Baconian guidelines for making natural histories, 'a thing ... of Excellent Political, and ... Œconomical Use' (Morton).[19] Placing his History of the Soil and Climate of South America right at the beginning of *A New Voyage*, that is, in that part of the book which otherwise neither deals with South America nor with scientifically precise observations, the author draws our attention to the second half

---

[18] For a fuller discussion of one of Defoe's favourite topics see chapter 8, pp. 173–4.
[19] The important link with Morton's Academy has been discussed above, pp. 70–1, and see p. 159, below.

of the book where this kind of information will be of central importance. Defoe's confident argument for colonisation is a product of the New Sciences, and the traveller-scientists' reports provided the model for applying the fundamental principles of the Baconian philosophy.

8

Defoe's *Tour*: a natural history of man and his activities

The previous chapter outlined the Royal Society's invitation to 'Seamen, Travellers, Tradesmen, and Merchants' to contribute with their observations to the making of a 'Natural History in general'. I have tried to show that Defoe was not only intimately familiar with the Fellows' directions for sea-bound voyagers but that he promoted these guidelines and so in a sense became a traveller-scientist by sea. Applying this framework of ideas in *A Tour thro' the Whole Island of Great Britain* (1724-6), Defoe composed his 'Natural History of a Nation'.

It is important to recall at this point that Bacon and his followers frequently employed the word 'history' when they referred to a systematic study of a set of natural phenomena devoid of theoretical speculation. Thus in the seventeenth and early eighteenth centuries 'history' could be used to describe either civil or natural history (and occasionally both; see p. 11 above). The notion of mixing the history of nature with that of nature improved 'by the works of man's hands' is well expressed in Robert Plot's *Natural History of Oxford-shire, being an Essay toward the Natural History of England* (London, 1677). In his preface 'To the Reader' Plot writes that his natural history tends '*not only to the advancement of a sort of Learning so much neglected in England, but of Trade also*' and he insists that:

> All which, without absurdity, may fall under the general notation of a *Natural History*, things of Art (as the Lord *Bacon* well observeth) not

> differing from those of Nature in *form* and *essence*, but in the *efficient* only; Man having no power over Nature, but in her matter and motion, i.e. to put together, separate, or fashion natural Bodies, and sometimes to alter their ordinary course.
>
> (Plot 1677: 1–2)[1]

The Baconian combination of histories of nature and of man will be of particular significance in the following discussion of Defoe's *Tour*. It is a point stressed in Boyle's advice to travellers by land. Following Bacon, Boyle divided all human knowledge into what 'respects the Heavens, or concerns the Air, the Water, or the Earth'. From these physical investigations, he advised that

> there must be a careful account given of the *Inhabitants* themselves, both *Natives* and *Strangers* ... in particular, their Stature, Shape, Colour, Features, Strength, Agility, Beauty (or the want of it), Complexions, Hair, Dyet, Inclinations, and Customs that seem not due to Education. As to their Women (besides the other things) may be observed their Fruitfulness or Barrenness; their hard or easy Labour, etc. And both in Women and Men must be taken notice of what diseases they are subject to, and in these whether there be any symptom, or any other Circumstance, that is unusual and remarkable.
>
> (*Philosophical Transactions* I: 188)

The members of the Society, as we have seen, were convinced that the voyager's first-hand 'Collecting, Preserving, and Sending over [of] Natural Things' was the best way to achieve the much hoped-for 'Universal Correspondence for the Advancement of Knowledge both Natural and Civil'. These words constitute part of the title of John Woodward's *Brief Instructions for Making Observations in all Parts of the World*

---

[1] A similar mixture of subject-matters is found in Moses Pitt's *The English Atlas* (Oxford, 1680–3). Sponsored by the Royal Society, Pitt's *Atlas* fulfilled the Fellows' demand for a comprehensive register, or history, of knowledge of fact. Pitt promised to give an account of every nation:

> their original Language, Manners, Religion, Employments, etc. that if any art or science useful to society be there eminent, it may be transferred into our own Country. Much more considerable are their Governments, Civil and Military, their Magistrates, Laws, Assemblies, Courts, Rewards and Punishments, and such like. Neither must we omit the manner of educating their youth in arts liberal and mechanick, taught in their Schools, Universities, Monasteries, Shops also, and the like. Their manner of providing for their poor of all sorts, either in Hospitals or Workhouses. Lastly, it will be expected, that we give an account of the History or actions and successes of each Nation, or their Princes, remarkable actions etc.
>
> (Pitt 1680–3, vol. I, Introduction, p. 4)

(London, 1696). As Woodward was particularly succinct in setting out the main Baconian principles for inquiring into the state of the country, he may be quoted here. Further, as his testimony reveals, many of the 'unusual and remarkable' things which the experimental scientists advised to be studied, found their way into Defoe's *Tour*.

Section III of Woodward's *Brief Instructions* directs the traveller in making observations 'At Land'. It is divided into ten principal headings. The first and second titles advise the traveller to compile a history of the '*Weather, Heat, Cold, Fogs, Mists, Snow, Hail, Rain, Thunder, Lightning, Meteors*, etc.'[2] Next come 'Observations concerning Springs':

> Let there be an account taken of all *Springs* ... Whether they contain *Bitumen, Petroleum, Salt, Nitre, Vitriol,* or other *Mineral Matter* in their Water. Upon what *Occasions,* or at what *Seasons* chiefly their Water *encreases* or *decreases.*

As to rivers, Woodward directs that their depth, breadth and quickness be observed; it should be noted what sorts of fish are found in the rivers and what plants grow along the shores. The traveller is next instructed to take particular notice of 'whether any Stones ... *resemble* the *Shells of Muscles, Cochles, Perewinkles,* or the like'. The fifth title concerns itself with 'Metalls, Minerals, Stones, Earths, etc.' The searcher into nature should also inquire into 'the *Damps*: of what *kind* they are: what *harm* they do: at what *season* chiefly they happen ...'. Detailed guidance is given on how to study 'Grottoes, and Mountains'. For example, it should be observed 'whether some of the *highest* of them have not their *Tops* covered with *Snow*, a great part, or all the Year'. The penultimate heading deals with 'the more observable and peculiar *Diseases* of the Country ... what *seasons* of the year are most subject to them: and of the other *Casualties,* particularly *Earthquakes,* noting all circumstances that *precede, attend,* and *follow after* them ...' The list closes with the advice for a history of plants and animals. In its detailed advice of what and how to observe and collect data, Woodward's *Brief Instructions*

---

[2] Woodward clearly responded to Bacon's advice for making histories, given in the *Parasceve.* In the 'Catalogue of Particular Histories by Titles' Bacon's tenth and eleventh titles consist of directions for the 'History of Showers, Ordinary, Stormy ...' and the 'History of Hail, Snow, Frost, Hoar-frost, Fog, Dew, and the like' (Bacon 1857, IV: 265 and cf. Hartlib's *Legacy*, 1651: 78–108).

admirably illustrates the close relationship between the Royal Society and the Restoration traveller by land (Woodward 1696: 3–8).

For the experimental scientists the gathering of facts was not an end in itself but a prerequisite for making agricultural, social and/or economic improvements. The compilation of up-to-date information, especially relating to social and economic investigation, was referred to as an inquiry into the 'Present State' of a country. We find the term in Edward Chamberlayne's *Present State of England* (London, 1669 and numerous reprints into the eighteenth century), Samuel Collins's *The Present State of Russia* (London 1671), William Petty's *The Fourth Part of the Present State of England* (London 1683)[3] – and we will encounter it once more in Defoe's *Tour*, which in the words of the author is 'A Description of "The present State" of England'. Invariably, the motivating force behind these scientific studies of the day was the hope that they might be inducive to the advancement and perfection not only 'of the Natural History of each Nation' but of husbandry, trade and commerce. For

> as it is evident, that except the benefits which God by Nature hath bestowed upon each Country bee known, there can be no Industrie used towards the improvement and Husbandry thereof; so except Husbandry be improved, the industrie of Trading, whereof a Nation is capable, can neither be advanced or profitably upheld.[4]

Defoe the traveller-scientist by land

The *Tour* opens with Defoe's assurance 'that he is very little in Debt to other Mens Labours, and gives but very few Accounts of Things, but what he has been an Eye-witness of himself'. He insists that his survey is neither 'the Produce of a cursory View, or rais'd upon the borrow'd Lights of other Observers'.[5] Similar guarantees of reliability are

---

[3] Other works that come into this category are Paul Rycaut, *The Present State of the Ottoman Empire* (London, 1668), Gerard Boate, *Ireland's Naturall History* (London, 1652), John Lawson, *The History of Carolina; Containing the Exact Description and Natural History of the Country: Together with the Present State thereof. And a Journal of a Thousand Miles, Travel'd thro' several Nations of Indians. Giving a particular Account of their Customs, Manners, &c* (London, 1714) compiled by Fred A. Olds and reprinted at Charlotte, North Carolina, 1903.

[4] From Hartlib's Epistle Dedicatory to Gerard Boate's *Ireland's Naturall History* (London, 1652).

[5] *A Tour thro' the Whole Island of Great Britain*, ed. G. D. H. Cole, 1927. All quotations follow this edition and are given in the text in the form I: 3.

liberally sprinkled through the text. If he could not personally make an observation, he stresses that his testimony is derived from persons 'of undoubted Credit, who [were] an Eye-Witness, and saw' it (I: 8, 13, 134, 149, 276, 290; II: 497, 520). 'Superficial Observers, must be superficial Writers' (I: 42) – Defoe is of a different calibre. He is concerned 'with the most exact Truth' (II: 520). More than once he writes that he has carried out recent research in order to update observations garnered years ago. Letter X begins

> Having thus finished my Account of the East Side of the North Division of England, I put a stop here, that I may observe the exact Course of my Travels; for as I do not write you these Letters from the Observations of one single Journey, so I describe Things as my Journies lead me, having no less than five times travelled through the North of *England*, and almost every time by a different Rout; purposely that I might see every thing that was to be seen, and, if possible, know every thing that is to be known, though not (at least till the last general Journey) knowing or resolving upon writing these Accounts to you.
>
> (II: 664)

There is abundant evidence that Defoe carried out specific research of at least some regions of England for the publication of the *Tour*.[6] He certainly portrays himself as the inquisitive, conscientious traveller who went from county to county, collecting and recording with an insuperable diligence information of the physical and man-made world (see p. 91 above).

Both the title-page and the Preface list the principal heads under which he intends to order his mass of 'useful Observations upon the Whole'. The *Tour* will concern itself with 'the Improvements in the Soil, the Product of the Earth, the Labour of the Poor, the Improve-

---

[6] In his *Review* for 22 February 1711 Defoe claimed:

> I have, within these 20 Years past, Travelled, I think I may say, to every Nook and Corner of that part of the Island call'd England, either upon Publick Affairs, when I had the Honour to serve his late Majesty King William, of Glorious (tho' forgotten) Memory – Or upon my private Affairs; I have been in every County, one excepted, and in every Considerable Town in every County, with very few Exceptions; I have not, I hope, been an idle Spectator, or a careless unobserving Passenger in any Place, and believe I can give some Account of my Travels if need were.
>
> (VII: 570 and V: 515)

(and see Moore 1958: 274–82). For Defoe's specific journeys for the publication of the *Tour*, see Rogers 1974–5: 431–50.

ment in Manufactures, in Merchandizes, in Navigation'. These observations of the man-made world (what the Baconians referred to as 'nature improved by the hand of man') will be combined with descriptions of the people, 'their Customs, Speech, Employments, the Product of their Labour, and the Manner of their living, the Circumstances as well as Situation of the Towns; their Trade and Government; of the Rarities of Art, or Nature; the Rivers, of the Inland, and River Navigation; also of the Lakes and Medicinal Springs' (I: 3). Defoe's main purpose is to inquire into 'the present State of Things', not as it has been but 'as it really is'. His intention of ascertaining 'the present State of Things' is repeated several times, in the Prefaces to volumes one and two and in the text (II: 541, 690, 691 and 703). In the Introduction to volume three the parallel with his seventeenth-century predecessors is made explicit:

> My Business is rather to give a true and impartial Description of the Place; a View of the Country, its present State as to Fertility, Commerce, Manufacture, and Product; with the Manners and Usages of the People ...
>
> (II: 541)

Clearly, Defoe's inquiry into the 'present State' of a country requires that the subject-matter of nature and man be combined.

With the watchful eye of the virtuoso, Defoe takes stock of the notable characteristics of the inhabitants of the various counties through which he passes. The inhabitants of Carmarthen are 'civiliz'd' and 'curteous' (II: 455); on reaching Cheshire, Defoe reflects that the '*Welsh Gentlemen* are very civil, hospitable, and kind; the People very obliging and conversible, and especially to Strangers ... willing to tell us every thing that belong'd to their Country, and to show us every thing that we desired to see' (II: 466). The people in Derbyshire are 'a rude boorish kind of People, but they are a bold, daring, and even desperate kind of Fellows in the Search into the Bowels of the Earth' (II: 565). Continuing his journey northwards, Defoe is determined

> not to quit *Northumberland* without taking notice, that the Natives of this Country, of the antient original Race or Families, are distinguished by a *Shibboleth* upon their Tongues, namely, a difficulty in pronouncing the Letter R, which they cannot deliver from their

Tongues without a hollow Jarring in the Throat, by which they are plainly known, as a Foreigner is, in pronouncing the Th.

(II: 662)

As is well known, one of the Society's fundamental principles was to report objectively, free from personal bias and prejudice. They reprehended writers who were

> more concern'd for Panegyricks of the amenities of the place, than will well sort with the true and modest relations of their Neighbours: As, when we read the beginning of the Ingenious Barclay's Euphormio, we are invited to prefer Scotland *before any Paradise on Earth; which yet I do not blame or censure in that noble Romance: But in our designed* Natural History *we have more need of severe, full and punctual Truth, than of Romances or Panegyricks.*
> (*Transactions* 1676–7, Preface to XI: 552)

In his 'Introduction to the Account and Description of *Scotland*' Defoe makes plain that he is not guilty of such bias: 'as I shall not make a Paradise of *Scotland*, so I assure you I shall not make a Wilderness of it' (II: 691). He writes that 'hitherto all the Descriptions of *Scotland*' have been subject to 'the most scandalous Partiality', what is needed is 'a more modern' and 'real Description' based on experience and 'critical Enquiries' (II: 689–90). Echoing the Society's aim at plain, 'punctual' (that is, precise or exact) truth, Defoe asserts: '*Scotland* is here describ'd with Brevity, but with Justice; and the present State of Things there, plac'd in as clear a Light as the Sheets ... will admit' (II: 690).

The author of the *Tour* takes account of the customs, manners, speech of the local people; in Boyle's words, Defoe gives 'a careful account of the *Inhabitants* themselves ... [their] Inclinations, and Customs that seem not due to Education'. Boyle had continued his advice for a search into the present state of a country: 'As to their Women ... may be observed their Fruitfulness or Barrenness ... what diseases they are subject to.' Is it coincidence that Defoe should write in this context of 'a strange Decay of the [female] Sex' caused by the 'unhealthy Marshes' near Tilbury in the Thames estuary?

> I was inform'd that in the Marshes on the other Side the River over-against *Candy* Island, there was a Farmer, who was then living with the five and Twentieth Wife, and his Son who was but about 35 Years old, had already had about fourteen.

As a warranty of truth, Defoe adds 'indeed this part of the Story, I only had by Report, tho' from good Hands too'. He rounds off his account with an appeal: 'to any impartial Enquiry, having myself Examin'd into it critically in several Places' (I: 13). Reading Defoe's observations of the appearance and distinctive characteristics of the inhabitants, and being aware of the Royal Society's *Directions* to travellers, it is hard not to be reminded of the above-quoted excerpt from Boyle's 'General Heads' or of Woodward's *Brief Instructions*, where the traveller is advised to 'take an account of the *Damps*: of what *kind* they are: what *harm* they do: at what *season* chiefly they happen' (see p. 153 above).

Throughout, Defoe asserts that his business is with the 'Matter of Fact'. Traditionally accepted 'wonders' are dismissed by reason and method. Arriving at the 'wonderful Place, the *Peak*', he will not 'do as some others have, (I think, foolishly) done before me, *viz.* tell you strange long Stories of Wonders' (II: 566). He discards Poole's Hole in Derbyshire as 'another *Wonderless* Wonder of the *Peak*' founded on nothing but 'antient Report':

> Let any Person therefore, who goes into Poole's Hole for the future, and has a mind to try the Experiment, take a long Pole in his hand, with a Cloth tied to the end of it, and mark any Place of the shining spangled Roof which his Pole will reach to; and then, wiping the drops of Water away, he shall see he will at once extinguish all those Glories; then let him sit still and wait a little, till, by the Nature of the thing, the Drops swell out again, and he shall find the Stars and Spangles rise again by degrees, here one, and there one, till they shine with the same Fraud, a mere *deceptio visus*, as they did before.

'In short,' Defoe concludes, 'there is nothing in *Poole's Hole* to make a Wonder of, any more than as other Things in Nature, which are rare to be seen, however easily accounted for, may be called wonderful' (II: 577–8). As for Weeden Well, which allegedly 'miraculously' fills and empties itself, Defoe explains it as 'a mere Accident in Nature':

> if any Person were to dig into the Place, and give vent to the Air, which fills the contracted Space within, they would soon see *Tideswell* turned into an ordinary running Stream, and a very little one too.
> (II: 580–1)

Defoe puts his faith in experience: 'Upon Experience 'tis found ...' (I: 58). Vindicating his belief in the method of the New Sciences, he insists on the value of personal observation. If a person will 'take a long

Pole in his hand', or 'dig into the Place' and will 'try the *Experiment* ... he shall *see*', and the wonder will soon be exposed as a 'Fraud' (my italics). Wonders based on 'antient Report', or, as Sprat had phrased it, 'the delightful deceit of *Fables*' which have 'been strengthned by long prescription', are dismissed since they cannot stand the test of evidence (Sprat 1959: 61–2).

Elden Hole, on the other hand, is acknowledged 'to be a Wonder': as ' Mr. *Cotton* says, he let down eight hundred Fathoms of Line into it, and that the Plummet drew still; so that, in a word, he sounded about a Mile perpendicular' (II: 584). The near bottomlessness of Elden Hole has been tested and proved to be true.[7]

A significant link here with Charles Morton occurs in Defoe's discussion of the 'Whispering Place' in Gloucester cathedral. By tradition, Defoe tells us, the 'Whispering Place' has 'past for something Miraculous'. His tutor had in fact discussed and disproved this 'miracle' in his science lectures, explaining to his students that 'nothing sets forth the truth of ... the Angles of Incidence and reflections in Sounds better than the famed Whispering place at Glosester which I have Seen, and observed'. Defoe's teacher supplemented his account with a discussion presented to the Royal Society on 5 November 1662.[8] It is a principal tenet of this thesis that Defoe's knowledge of the Royal Society, its goals and methodology, came through his contact with Charles Morton. In this context it is noteworthy that Morton should have first inquired personally into the 'Whispering Place' in Gloucester cathedral and then tested his experience with the Baconian method of investigation, thus rejecting the reputed mystery. Now, decades later, Defoe travels to Gloucester to examine the same phenomenon and similarly rejects it: 'since experience has taught us the easily comprehended Reason of the Thing: And since there is now the like in the Church of St. *Pauls*, the Wonder is much abated' (II: 440). It is significant, too, that Defoe's refutation of the 'miraculous' through experience, is carried out in the words in which he had a few

---

[7] As will be shown in detail below, many of Defoe's confident assertions are based on scientific investigations given in Gibson's edition of Camden (this one can be found in col. 498). Still within Defoe's life-time the wonder of Elden Hole was dispelled when a Fellow of the Royal Society found the depth to be no more than 76 feet: see *Tour*, ed. Rogers 1971, note 24 to Letter VIII.

[8] Morton 1940: 167; Birch 1756, I: 120–3 and see p. 40 above.

years earlier made Crusoe reject the 'miraculous' growth of a grain of corn (*Crusoe* 1, p. 78).

Defoe's use of 'unquestionable testimony'

In *The Great Law of Subordination Consider'd*, also published in 1724, Defoe gives an indication of how he gathered reliable material for his *Tour*:

> As thus I made myself Master of the History, and *ancient State* of *England*, I resolv'd in the next Place, to make myself Master of *its present State* also; and to this Purpose, I travell'd in three or four several Tours, over the whole Island, critically observing, and carefully informing myself of every thing worth observing in all the Towns and Countries through which I pass'd.
>
> I took with me an *ancient Gentleman* of my Acquaintance, who I found was thorowly acquainted with almost every Part of *England*, and who was to me as a walking Library, or a moveable Map of the Countries and Towns through which we pass'd; and we never fail'd to enquire of the most proper Persons in every Place where we came, what was to be seen? what Rarities of Nature, Antiquities, ancient Buildings were in the respective Parts? or, in short, every thing worth the Observation of Travellers.
>
> (pp. 46–7)

Apart from this '*ancient Gentleman*', if he ever existed, there is, however, another source which proves to be Defoe's most consistent 'walking Library' and 'moveable Map', namely, Camden's *Britannia* in the edition by Edmund Gibson (London 1695).[9]

When Gibson (1669–1748), then of Queen's College, Oxford, subsequently Bishop of London, agreed to edit *Britannia*, it was decided that the account of each county should be updated with recent research. A scholarly team was appointed to help with the undertaking. Listing these collaborators, the Preface mentions among others John Evelyn, Robert Plot (who carried out a special 'survey of *Kent* and *Middlesex* ... upon this occasion') and Edward Lhwyd, who succeeded Plot as Keeper of the Ashmolean Museum. Gibson wrote of this contributor: 'When I tell you, that the whole business of *Wales* was

---

[9] This edition was reprinted in two volumes in 1722. Since the single-volume edition of 1695 is listed in the Sales Catalogue of the Defoe/Farewell libraries, it seems safe to assume that Defoe uses the 1695 text: see Heidenreich 1970. Citations from *Britannia* are to the 1971 facsimile reprint.

committed to the care of Mr. *Edward Lhwyd* ... no one ought to dispute the justness and accuracy of the Observations.' From John Aubrey, another devoted Baconian, Gibson managed to get permission to use extracts from his unpublished *Monumenta Britannica*. Another keen traveller-scientist, Ralph Thoresby, was responsible for the updated notes for West Riding in Yorkshire. John Ray, undoubtedly the greatest Baconian naturalist of the century, was appointed to provide the 'Catalogue of Plants' which was now added at the end of each county. Thus, a considerable number of the Fellows of the Royal Society contributed to Gibson's new edition. True to their belief in studying nature directly, they carried out special journeys to furnish Gibson with up-to-date research. They also frequently used and referred to each other's work. Reading the modern sections added by Gibson, we are constantly reminded of the experimental scientists' collaborative approach to knowledge – if Defoe did not know before, and at first hand, of Ray, Plot, Aubrey, Lhwyd, Childrey and Woodward, he certainly knew of them after his careful study of Gibson's edition of Camden. It hardly needs stressing that Gibson's scholarly team altered Camden's original focus on antiquities and ecclesiastical foundations, supplying information on natural resources of each region, the inhabitants, their customs, manner, and trade; this is the information that particularly interested Defoe.

Defoe's *Tour* refers to either Camden or Gibson about eighty times by name. In addition to these direct references and citations, there are many more oblique ones where Defoe quotes, paraphrases or adapts without mentioning his source. Except for Letter V dealing with London, his use of 'Camden and his Right Reverend Continuator' is ubiquitous. It appears that the less Defoe knew an area, the more he relied on Camden/Gibson.[10]

His greatest indebtedness occurs in Letters VIII, IX, and X, dealing with Yorkshire, Derbyshire, and Lancashire. For instances

---

[10] For the sources for Defoe's *Tour* see Davies 1950, Andrews 1959, Bastian 1967, Rogers 1973 and the Introduction to Rogers' edition of the *Tour* 1971. In 1967 Frank Bastian wrote: 'One of [Defoe's] mainstays was Camden's *Britannia*, in the 1695 edition of Edmund Gibson, with important "Additions" by various hands', and six years later Pat Rogers noted 'an extremely heavy levy upon *Britannia*, and specifically the edition of 1695 by Edmund Gibson'. No commentator on Defoe, as far as I could discover, has taken into account that Gibson was helped by a considerable number of Fellows of the Royal Society.

where he lifts his text virtually unchanged from the modern 'Additions', the following may serve as examples. He writes knowingly of the magistrates of Halifax that their judgement is passed only if a thief is caught

> 1. *Hand Napping*, that is, to be taken in the very Fact, or, as the Scots call it in the Case of Murther, *Red Hand*.
> 2. *Back Bearing*, that is, when the Cloth was found on the Person carrying it off.
> 3. *Tongue Confessing*, that part needs no farther Explanation.
>
> (II: 608)

Here is the relevant passage from Gibson:

> the fact must be certain; for he must either be taken *hand-habend*, i.e. having his hand in, or being in the very act of stealing; or *back berond*, i.e. having the thing stoln either upon his back, or somewhere about him, without giving any probable account how he came by it; or lastly *confesson'd* [*sic*], owning that he stole the thing for which he was accused.
>
> ('Additions' col. 726)

Continuing in Yorkshire, Defoe writes that he 'saw *Knaresborough*' and its four 'magical' springs. Dismissing these 'wonderless wonders' as something natural and explicable, he goes on to describe two of them.

> 1. The first is the *Sweet Spaw*, or a Vitriolick Water; it was discovered by one Mr. *Slingsby*, *Anno* 1630. and all Physicians acknowledge it to be a very sovereign Medicine in several particular Distempers. *Vid. Dr. Leigh's Nat. Hist. of Lancashire.*
> 2. The *Stinking Spaw*, or, if you will, according to the Learned, thc *Sulphur Well*. This Water is clear as Chrystal, but fœtid and nauseous to the smell, so that those who drink it are obliged to hold their Noses when they drink; yet it is a valuable Medicine also in Scorbutic, Hypochondriac, and especially in Hydropic Distempers; as to its curing the Gout, I take that, *as in other Cases, ad referendum.*
>
> (II: 619)

This apparent first-hand account ('we saw *Knaresborough*') comes straight from the 'Additions', where we read that this town is chiefly famous for its four medicinal springs:

1. The *Sweet-spaw* or *Vitrioline-well*, discover'd by Mr. *Slingsby* about the year 1620. 2. The *Stinking* or *Sulphur-well*, said to cure the Dropsie, Spleen, Scurvy, Gout, etc. so that what formerly was call'd *Dedecus Medicinae*, may be call'd *Decus Fontis Knaresburgensis*, the late way of *bathing* being esteem'd very soveraign.

('Additions' cols. 732–3)[11]

Since Gibson/Camden fail to describe the third and fourth springs, it is not surprising that Defoe also omits them. One wonders why Defoe felt constrained to make minor changes to his borrowings (as to the spring's curing of the gout, moving the date from 1620 to 1630 – curiously enough, he has no qualms about re-using the word 'soveraign'). Did his private observations not square with that of the scholars? Similar slight adaptations occur in his description of the well near Scarborough:

> It is hard to describe the Taste of the Waters; they are apparently ting'd with a Collection of Mineral Salts, as of Vitriol, Allom, Iron, and perhaps Sulphur, and taste evidently of the Allom.
>
> (II: 656)

Gibson's scientific team of investigators had reported of the 'Spaw-well' near Scarborough:

> It's virtue proceeds from a participation of *Vitriol, Iron, Alum, Nitre* and *Salt*: to the sight it is very transparent, inclining somewhat to a sky-colour: it hath a pleasant acid taste from the *Vitriol*, and an inky smell.
>
> ('Additions' col.765)

We recall that the Fellows of the Royal Society had instructed that 'there be an account taken of all *Springs* ... Whether they contain *Bitumen, Petroleum, Salt, Nitre, Vitriol*, or other *Mineral Matter* in their Water' (Woodward 1696: 4 and see p. 153 above). These are the scientific data that the virtuosi supplied for the revised edition of *Britannia*; in turn, these are the details which Defoe takes from Gibson's Camden, and then (presumably) tests and corrects against his own experience in order to make his account as truthful and up to date as possible.

The most concentrated unacknowledged 'cribbing' comes in his

---

[11] See also Defoe's mention of the 'Stories of the Well of S. Winifred' which he dismisses: they 'smell too much of the Legend' (II: 464) – his account was based on that given in the 'Additions' by Gibson, cols. 690–1.

description of Beverley. Defoe here unashamedly plagiarises ten paragraphs verbatim ('Additions' cols. 743–4; *Tour* II: 644–7).[12] But this is an exception. His usual method of borrowing is more selective and skilful. He takes the basic facts, interweaves them with his own knowledge, and so composes a contemporary 'history' of England that is alive with a superabundance of detail. 'In Hull', he records:

> They shew us still in their Town-Hall the Figure of a Northern Fisherman, supposed to be of *Greenland* ... He was taken up at Sea in a Leather Boat, which he sate in, and was covered with Skins, which drew together about his Waste, so that the Boat could not fill, and he could not sink; the Creature would never feed nor speak, and so died.
>
> (II: 653)

Defoe may have been shown the figure of a native of Greenland at Hull, but the basic details are derived from the 'Additions'. In this case the 'accurate description' had been supplied by 'the curious and ingenious Mr. *Ray*' who, the editor tells us, had 'actually view'd' the town-hall of Hull. Gibson quotes at length from Ray's eye-witness account; I shall re-quote part of the relevant passage to show what Defoe had read, what he selected, and how he transformed the material. Ray had reported:

> In the midst of this room hangs the effigies of a native of *Groenland*, with a loose skin-coat upon him, sitting in a small boat or *Canoe* cover'd with skins; and having his lower part under deck. For the boat is deck'd or cover'd above with the same whereof it is made, having only a round hole fitted to his body, through which he puts down his legs and lower parts into the boat ... The Groenlander that was taken refused to eat, and died within three days after.
>
> ('Additions' col. 745)

It is not often that we can look over Defoe's shoulder and see him at work. Gibson's Camden offers this rare opportunity. Here we witness what Defoe inherited and how he inserted scientifically collected information into his 'first-hand' account.

A much-debated question in the second half of the seventeenth

---

[12] And see Andrews 1959: 339 and Rogers 1973: 773. Defoe's only addition is the word 'present' in the seventh paragraph (*Tour* II: 646). Defoe's sentence reads: 'But to come to the present Condition of the Town ...', the author's main aim of inquiring into 'the present State of Things' is once more affirmed.

century was '*whether the stones we find in the forms of* Shell-fish, *be* Lapides sui generis, *naturally produced . . . Or whether they rather owe their form and figuration to the shells of the* Fishes *they represent*' (Plot 1677: 111). Camden, in his account, written in the Tudor age, maintained that the snake-stones were 'strange frolicks of nature, which . . . she forms for diversion after a toilsome application to serious business'. A century and a half later Gibson's contributor commented that the question of fossil-stones was as yet unresolved and 'has been very much controverted by several Learned men on both sides'. Gibson's 'Additions' directed the reader to the research of Lhwyd, Beaumont, Ray and Woodward – Woodward was by Defoe's time *the* leading scientist in this field (cols. 751 and 765; a complete list would have also included Hooke and Plot). Defoe may have followed Gibson's advice and checked these authors before discussing the fossils found near Musgrave: 'Next here are the Snake Stones, of which nothing can be said but as one observes of them, to see how Nature sports her self to amuse us, as if Snakes could grow in those Stones' (II: 657). His use of the current phrase at the time 'to see how Nature sports her self' would indicate that he was familiar not only with Gibson/Camden but with the debate as such. If his opinion strikes us today as un-modern, then we must remember that even Dr Plot thought that fossil-stones were 'lusus naturae' (sports of nature), and that Ray by his own statement 'fluctuated for a long time in [his] opinion concerning the Originall of these Stones' (Levine 1977: 26) – Hooke, of course, rejected it in his Cutlerian Lectures to the Royal Society.

From the evidence collected so far it can be said that there can be no doubt that when Defoe composed his text of Northern England, he had Gibson's Camden in front of him and scanned its pages for reliable data. When he writes 'we saw', 'it is observ'd', 'they show us still', 'it is hard to describe', 'I enquired into', sustaining the general impression of a personal traveller's account, he frequently bases his facts on the sections collected by modern – that is to say, scientific – methods of research. Defoe adds to his private observations such information as seems to him incontrovertible.

In the *Tour* reliable second-hand material and personal knowledge are so 'artfully' welded together that it is impossible to say with certainty where one ends and the other begins. Not unlike the Fellows of the Royal Society, Defoe relies on those who have 'actually review'd the evidence'. However, there is a fundamental difference: while Baconian experimental scientists, characteristic of their belief in collaborative efforts, sought to add authenticity and value by citing the source of their information ('as Dr. Woodward stated'; 'as the curious and ingenious Mr. Ray reported'), Defoe adopts the virtuosi's research without acknowledgement. He uses their 'faithful reports' only to conceal them under a cloak of common knowledge which 'may be learn'd from due Enquiry and from Conversation'. Defoe shows himself to be the professional reporter rather than the scrupulous auxiliary scientist. When he notes that 'the Country People told us a long Story here of Gipsies', or '*Vipseys*', and goes on to inform us that these water-spouts are 'really natural *Jette d'eaus* or Fountains' (II: 655), it appears as if this information was gathered from local gossip, when in fact it is based on the work of Ray and Childrey. In Gibson's 'Additions' we read:

> Concerning the *Vipseys* hereabouts, take what the ingenious Mr. Ray was pleas'd to communicate, among other things relating to these parts. 'These *Vipseys*, or suddain eruptions of water, whether the word in *Newbrigensis* were by mistake of the Scribe, and change of a letter, put in stead of *Gipseys*; or whether *Vipseys* were the original name, and in process of time chang'd into *Gipseys*, I know not; certain it is they are at this day call'd *Gipseys* ... Neither are these eruptions of Springs proper and peculiar to the wolds of this Country, but common to others also, as Dr. *Childrey* in his *Britannica Baconica* witnesseth.'
>
> (col. 748)

Defoe finishes his account with: 'That which was most observable to us, was, that the Country People have a Notion that whenever those *Gipsies*, or, as some call 'em, *Vipseys*, break out, there will certainly ensue either Famine or Plague' (II: 655). We will, of course, never know for certain how much Defoe gleaned from personal conversation and how much he took from Childrey's *Britannica Baconica*, which reported that the *Gipseys* were 'reputed by the common people a forerunner of dearth' (col. 748). What mattered to Defoe, and what he believed mattered to posterity, was not the source of his information but that the *Tour* should be a reliable repository of observations drawn from and verified by life.

An interesting case in point comes in Letter I, Defoe's 'Account of a petrifying Quality in the Earth' near Harwich.

> The Fact is indeed true, for there is a sort of Clay in the Cliff ... which when it falls down into the Sea, where it is beaten with the Waves and the Weather, turns gradually into Stone: but the chief Reason assign'd, is from the Water of a certain Spring or Well, which rising in the said Cliff, runs down into the Sea among those pieces of Clay, and petrifies them as it runs ... The same Spring is said to turn Wood into Iron: But this I take to be no more or less than the Quality.
>
> (I: 35–6)

Defoe's argument is that appearances are deceptive, and that the clay is neither turned into stone nor the wood turned into iron.

> I presume, that those who call the harden'd Pieces of Wood, which they take out of this Well by the Name of Iron, *never try'd* the Quality of it with the Fire or Hammer; if they had, perhaps they would have given some other Account of it.
>
> (I: 36; my italics)

Behind Defoe's well-argued case for personal experience stand the experimental scientists' discussions on petrification of earth and wood as we find them in the 'Additions' (col. 359; see Sprat 1959: 255 and see pp. 85–6 above). He may well speak with assuredness, since his account is founded upon the then available facts established by the highest authority.

Defoe knew that to be a 'modern' required constant questioning and testing, even of the experts. He never misses an opportunity to point out his own superior knowledge (I: 396; II: 440, 455–6, 589, 752, 824 etc.). It is with special satisfaction that he dismisses both Camden's and Gibson's (or rather Evelyn's) report of the disappearing river Mole near Box Hill in Surrey. According to their description the Mole is '*Swallow'd up*' by the earth. Defoe, who is on home-ground in Surrey (since he 'liv'd in the Neighbourhood several Years'), rides into the attack:

> 'Tis strange this Error should prevail in this manner, and with Men of Learning too, and in a Case so easily discover'd and so near. But thus it is, nor is it at all remote from the true design of this Work, to undeceive the World in the false or mistaken Accounts, which other

> Men have given of Things, especially when those mistakes are so demonstrably gross.
>
> (I: 148)

The truth is that the waters of the Mole are here dispersed, 'so that there is no such thing as a ... River lost, no, not at all'. And this, Defoe concludes with his habitual stress on experience, 'I affirm of my own knowledge, having seen it so, on many Occasions' (*Tour* I: 149; Gibson/Camden cols. 156 and 163).

The *Tour* provides us with unquestionable evidence of Defoe's conscious alignment with the Baconian method of inquiry. His concealed method of borrowing has for centuries succeeded in hiding the origin of his information. However, once we know the source of his scientific information and see his imaginative intertwining of personal and second-hand experience, we can be warned to evaluate his views with care. He may be expressing a personal opinion, or he may be borrowing, or at least supplementing a private theory with the work of those whom he esteemed reliable. One thing is certain, it is not by chance that Defoe takes account of water-spouts, meteors or 'livid fires', legendary wells (how they increase and decrease and what minerals they contain), snake-stones, the petrification of earth and wood, the nature of sound, etc.[13] First introduced to these and related subjects at Morton's Academy, Defoe retained a life-long and informed interest in a scientific approach to phenomena (cf. the *Compendium Physicae* where Morton explored these questions, in Morton 1940: 72, 81ff, 99ff, 125, etc.; and see pp. 39–42 above). If Defoe discusses these subjects in the *Tour* he does so not because he happens to stumble across this information in Gibson's Camden, but because he genuinely believes that this kind of precise, scientific knowledge would make his report complete and useful. Again, it is no coincidence that he should combine these natural observations with the study of the customs, manners,

---

[13] For his discussion of 'a livid Fire' (*Tour* II: 460) Defoe refers us to the *Philosophical Transactions*. A letter from a 'Mr. *Maurice Jones*' giving 'An Account of the Burning of several Hay-Ricks by a Fiery Exhalation or Damp' was printed in 1694, XVIII: 49–50. This account is also printed in Gibson's Camden, cols. 659–60. In a 'farther Account of the Fiery Exhalation' in the *Transactions*, XVIII: 223, it appears that Edward Lhwyd was the recipient of Jones's letter. The above account of Defoe's use of Gibson's 'Additions' concentrates on three letters and gives but an idea of his indebtedness. A full discussion is left to my article 'Defoe the traveller-scientist: the use of Gibson's Camden in making *A Tour thro' the Whole Island of Great Britain*' (forthcoming).

speech, trade of the inhabitants; the combined history of nature and of man stood at the centre of the system of thought which he shared.

In the *Tour* the Baconian study of physical reality and the traveller's report merged into one, indeed, in the process of merging the two categories a curious reversal of roles can be observed. While the seventeenth-century traveller contributed to the Society's transactions, and was proud of the unprecedented status he attained as amateur scientist, Defoe raids their stock and conceals his booty. He refuses to be placed on the same level as the auxiliary scholar. Writing not for the intelligentsia who expect scholarly cross-references and acknowledgements, but for practical men 'of the most unaffect'd, and most unartificial kinds of life', Defoe uses different standards. He purloins knowledge from here and there, blends it and creates a trustworthy and, above all, useful account based on '*severe, full and punctual Truth*'. However, in choosing to use the New Scientists' faithful reports and to incorporate them into his own, he affirms his belief in the Baconian philosophy and becomes a promoter of 'useful knowledge both natural and civil'.

'*To Propagate useful Knowledge, for the good of Mankind*'

Defoe is convinced that 'no Man can do his Country a greater Service, than to open their Eyes, and encourage their Hands to Industry and Improvement' (*History of Trade* III: 43). One of the leading ideas of the *Tour* is to open British eyes and 'to let them see' how they are favoured by God with a superabundance of natural resources. Defoe is deeply committed to the Baconian belief that man is privileged '*above other Creatures* [in] *that we are not only able to* behold *the works of Nature, or barely to sustein our lives by them, but we have also the power of* considering, comparing, altering, assisting, *and* improving *them to various uses*' (Hooke 1961, Preface). His self-appointed duty is to make his countrymen aware of their God-given prerogative and to call them to industry and diligence. The purpose of writing, he states in his history of science, published at the same time as the *Tour*, is to '*encourage from the success of former Times to pursue the like useful Discoveries for the benefit of the Ages to come*'. As has already been pointed out, Defoe avows in this work '*the same Zeal for the general improvement of the World*' which inspired previous generations; it prompts him to '*Propagate useful Knowledge, for the good of Mankind*' (*History of Arts and Sciences*, Preface and see p. 79 and 98 above). The truth is, nearly everything Defoe wrote, and particularly the *Tour*, is permeated with the Baconian and Restoration voyager's 'hearty Zeal for the promoting of useful Knowledge, and of any thing that may never so remotely tend to [his] Countries Advantage' (Dampier's phrase, see p. 135 above).

When Defoe surveys a country and its people, he does so from the point of view of their usefulness. The question uppermost in his mind is, what has man done (or failed to do) to alter, assist and improve nature for his benefit and use? Rivers are studied as convenient 'Chanels for conveying an infinite Quantity of Provisions', they are links between towns or the city and the country. Defoe will 'sing you no Songs here of the River [Thames] in the first Person of a Water Nymph, a Goddess, (and I know not what) according to the Humour of the ancient Poets',[14] instead he will speak of it 'as it really is *made glorious*' by the hand of man:

> gilded with noble Palaces, strong Fortifications, large Hospitals, and publick Buildings; with the greatest Bridge, and the greatest City in the World, made famous by the Opulence of its Merchants, the Encrease and Extensiveness of its Commerce; by its invincible Navies, and by the innumerable Fleets of Ships sailing upon it, to and from all Parts of the World.
>
> (*Tour* I: 173–4)

Plains, meadows, forests are treated from the possibility of agrarian prosperity. Travelling in the New Forest, Defoe invites the reader to 'observe these Hills and Plains'. He describes them as being 'most beautifully Intersected, and cut thro' by the Course of divers pleasant and profitable Rivers ... there always is a Chain of fruitful Meadows, and rich Pastures' (I: 192). Beholding nature, Defoe is rarely moved to poetic appreciation; instead he reflects what resources and opportunities there might be, to render it more 'fruitful', 'profitable', and 'rich'.

Not only does Defoe industriously collect and record details about the natural and man-made world, but he shares the New Scientists' conviction that accurate information should form the basis for 'the needful Account of Alterations and Improvements' (II: 542). As the

[14] For Camden's poem to 'Father *Thames*' see cols. 147–8 and see Rogers 1980: 283–99.

traveller-scientist Defoe resolves 'to have a perfect Knowledge of the most remarkable Things', concentrating his attention on 'Manufactures', that is, on nature altered by the activities of men. 'I cannot believe,' he writes, 'that God ever design'd the Riches of the World to be useless to the World' (*History of Arts and Sciences*, p. 6). In the *Tour* this characteristic Baconian idea is formulated in these terms:

> I cannot think, but that Providence, which made nothing in vain, cannot have reserv'd so useful, so convenient a Port to lie vacant in the World, but that the Time will some time or other come (especially considering the improving Temper of the present Age) when some peculiar beneficial Business may be found out, to make the Port of Ipswich as useful to the World, and the Town as flourishing, as Nature has made it proper and capable to be.
> (I: 44–5)

Stressing his position as propagator of useful knowledge, Defoe comments: 'What I have said, is only to let the World see, what Improvements this Town and Port is capable of.' As we have seen above in the discussion of Defoe's *History of Trade*, the gathering of reliable information is intended principally to make suggestions for future improvements (see pp. 84–5 and 97–8 above).

Diligence, wisdom and experience appear to ensure prosperity of every kind. This is not so. Defoe is keenly aware of both the power and the frailty of human endeavour. The man-made world, he argues, is yet part of God's creation and subject to laws beyond our control. Towns, cities, trade remain part of nature, are nature improved by the 'Works of Mens Hands' (*History of Trade* III: 5). Rather than the traditional classical and Renaissance polarisation of art and nature, in Defoe art is most often seen as the continuity or adaptation of nature. The Baconian scientists, we recall, were careful to point out that their 'faithful *Records*' were of '*Nature, or Art*'; they explained that their studies of trades and craft-techniques were histories of nature improved by the activities of men. This on the whole is Defoe's point of view. 'Towns, Kings, Countries, Families', although made by man, follow 'the Course of Nature', they grow and flourish for a brief time, then they decline and die, and this is due to no other reason but 'their Destruction in the Womb of Time' (I: 54). The declining town of Southampton is 'in a manner dying with Age' (I: 141), and Dunwich stands as such another 'Testimony of the decay of Publick Things, Things of the most durable Nature'. For this town to 'Decay, as it were of itself (for we never read of Dunwich being Plundered, or Ruin'd, by any Disaster, at least not of late Years); this I must confess, seems owing to nothing but to the Fate of Things' (I: 54).

A natural history of man and his activities, the *Tour* reminds us of man's responsibility to study and master nature; we are also made aware that man is not always in charge, and that not all decay is due to his neglect. Take Ayr: it was 'formerly a large City, had a good Harbour, and a great Trade' but is now fallen into decay. Defoe compares Ayr to 'an old Beauty, [which] shews the Ruins of a good Face', stressing thereby the natural life-span of the town. 'The Reason of its Decay, is, the Decay of its Trade', but 'what the Reason of the Decay of Trade here was, or when it first began to decay, is hard to determine' (II: 739–40). Defoe sees decay as a natural process happening 'as it were of it self'. Forces beyond our determination or knowledge attack the safest stronghold and the soundest business. No doubt a personal note is struck when he comments that 'Wealthy City Families' have considered 'their Houses establish'd' when unexplainable 'Misfortunes of Business, and the Disasters of Trade' have destroyed them, and this at a time when 'the World has thought them pass'd all possibility of Danger' (I: 169).

Although Defoe's declared aim is to give 'the present State' of England, he is drawn into investigating what causes the decay and 'the History of pass'd Ages'. Almost against the author's will, the work combines the study of nature and man with comments on antiquity. Hooke, Aubrey, Woodward, Plot, Stukeley, Sloane and others demonstrated in their work that the study of the natural and man-made past were closely related. Making the analogy between natural and antiquarian knowledge explicit, Hooke had written: 'There is no Coin can so well inform an Antiquary ... as these [fossil shells] will a Natural Antiquary ... methinks Providence does seem to have design'd these permanent shapes, as Monuments and Records to instruct succeeding Ages of what past in preceding' (Hooke 1705: 321). The subjects link up naturally not only because Art emerges from Nature but because they share the same laws: 'Time [is] the great Devourer' of both nature and the works of man (II: 447, 473). Defoe reflects that

> though the Earth, which naturally eats into the strongest Stones, Metals, or whatever Substance, simple or compound, is or can be by Art or Nature prepared to endure it, has defaced the Surface, the Figures and Inscriptions upon most of these Things, yet they are beautiful, even in their decay.
>
> (II: 663)

His sensitivity to the limits of human power is both poignant and reassuring. Man can do so much to prepare and adapt nature against the onslaught of time, yet in the end he is not the master but the 'viceroy to the King of all the earth'.[15]

He is dismayed when he discovers that people have wilfully ignored their God-given duty to improve nature. As early as 1709 he devoted an issue of the *Review* to 'an Enquiry into the Methods for improving the Lands of *Scotland*, as being the only Foundation-Step of raising the present Circumstances of that injur'd Country'. Calling the Scots to alacrity, he urges: 'don't lay your own Sloth upon the Back of your Country, and load your Maker with what is meerly your own Fault' (*Review* VI: 190–1). Similarly in *Caledonia* (1706), Defoe follows his hymn to the praise of science with his encouragement to the Scots to bestir themselves:

> Wake *Scotland*, from thy long *Lethargick* Dream,
> Seem what *thou art*, and be what thou shalt seem.
>
> (p. 54)

More than twenty years later Defoe repeats these ideas, his *An Humble Proposal to the People of England* (1729), being written '*as one Alarm more given to the lethargick Age, if possible to open their Eyes to their own Prosperity*'. He encourages his countrymen to industry since they 'have not only the means of Improvement in [their] hands, but the Capacity of improving also'. Presenting a vision of hope and great national power, Defoe writes:

> Let us see in a few Words what Nature and Providence has done for us; nay, what they have done for us exclusive of the rest of the World. The Bounty of Heaven has stor'd us with the Principles of Commerce, fruitful of a vast variety of Things essential to Trade, and which call upon us as it were in the Voice of Nature, bidding us work, and with annex'd Encouragement to do so from the visible

---

[15] From *Crusoe* 3, p. 179; see also pp. 114–19 above, where this point is made in greater detail.

> apparent Success of Industry. Here the Voice of the World, is plain like the Answer of an Oracle, thus *Dig* and *Find*, *Plow* and *Reap*, *Fish* and *Take*, *Spin* and *Live*, in a word, *Trade* and *Thrive*.
> (Preface and pp. 1–9 and cf. *A Brief Deduction of the Original, Progress and Immense Greatness of the British Woollen Manufacture* (1727))

The same framework of ideas supports the *Tour*. Journeying through Scotland he is appalled by the general ignorance and laziness he observes.

> Here is a pleasant Situation, and yet nothing pleasant to be seen. Here is a Harbour without Ships, a Port without Trade, a Fishery without Nets, a People without Business; and, that which is worse than all, [the people of Kirkcudbright] do not seem to desire Business, much less do they understand it.
> (II: 733)

Because the Scots have been '*contented with such Things as they have*' and have not exerted themselves to improve nature, they have sinned against both God and man. 'They have all the Materials for Trade, but no Genius to it; all the Opportunities for Trade, but no Inclination to it.' Defoe's comment puts us in mind of Sprat's eulogy of the English 'merchant-voyager', who is inspired with a '*Noble*, and *Inquisitive Genius*' (Sprat 1959: 88). In Scotland there is the possibility of 'extraordinary Salmon Fishing, the Salmon come and offer themselves, and go again, and cannot obtain the Privilege of being made useful to Mankind; for they take very few of them'. Defoe writes that the Scots

> have all the Invitations to Trade that Nature can give them, but they take no Notice of it. A Man might say of them, that they have the *Indies* at their Door, and will not Dip into the Wealth of them; a Gold Mine at their Door, and will not Dig it.
> (*Tour* II: 733–4 and see *History of Trade*, III: 34–5)

He insists, not just once but many times, that man must follow Solomon (the first natural historian) and '*Search for knowledge as for silver and dig for it as for hid treasure*' (Defoe's quotation in *Gentleman*, p. 37, is from Proverbs 2: 4).

It is reasonable to judge the importance of these ideas in Defoe's thinking by the frequency with which they occur. First formulated in print in 1706 in *Caledonia*, this theme is reiterated in the *History of Trade* (1713), *History of Arts and Sciences* (1725–7), *Crusoe* 3 (1720) and in *A New*

Voyage (1724). After its re-appearance in the *Tour*, Defoe employs it once more in two works written towards the end of his life, in *Atlas Maritimus and Commercialis* (1728) and in the *Compleat English Gentleman*. He invariably portrays the 'Merchant-Adventurer' as one who is determined to find out 'what lies treasur'd up in the Bowels of the Earth, or in the Remote Parts of Uninhabited Climates, and Unnavigated Seas, Bays, Channels, and Retreats of the Waters' (*Review* IX: 108 and *Brief Observations on Trade and Manufactures* (1721), p. 5). Convinced of the quintessential Baconian belief that 'the Treasures of Nature are conceal'd, as Rareties inaccessible but by Labour, reserv'd as a Reward to the Industrious, and deny'd to the Slothful as a just Punishment of their Sloth', Defoe applies himself to uncover useful information (*Review* VI: 191). He is diligent 'to see every thing that was to be seen, and, if possible, know every thing that is to be known' (*Tour* II: 664). When, not long after the publication of the *Tour*, Defoe re-uses these words, he puts them into the context in which he wishes them to be understood. Drawing the parallel between Solomon, the Baconian prototype, and the 'real' scholar, he states that he himself

> abhorr'd to be ignorant of any thing, and from hence he resolv'd to see every thing that was to be seen, hear every thing that was to be heard, know every thing that was to be known and learn every thing that was to be taught.
> (*Gentleman*, p. 36 and see pp. 15, 60 and 116 above)

Defoe defines himself. Not by anybody else's statement but by his own, is he the Baconian 'strict Search[er] into every thing that is curious in Nature' (*Tour* II: 663). Committed to useful knowledge, he is resolute in 'improving himself that he might improve his whole empire' (*Gentleman*, p. 37), or, as he expressed it in another work written at this time, he resolved 'upon a general Application to the great Work of informing himself, and by degrees his People also in the useful Knowledge'.[16] If Defoe has persuaded generations of readers to see the old world through modern eyes, it is not only because Bacon or the Fellows of the Royal Society have thought that sceptical, first-hand investigations were vitally important in mapping the present state of things, but

---

[16] *An Impartial History of the Life and Actions of Peter Alexowitz, the Present Czar of Muscovy* (1723), p. 6.

because he, by his own experiment, found these principles to be true.[17] 'Experience has taught [him] the easily comprehended Reason of the Thing' (II: 440). The *Tour* vindicates the belief in the promotion of useful knowledge; a comprehensive view of the whole is considered possible and useful because personal observations and personal experience of a life-time have dictated it to be so. Our thought and vision are guided by the essential tenets of Bacon's philosophy, yet we perceive Defoe's England as it in fact was.

[17] Cf. Trevelyan 1946. Trevelyan, who entitled his chapter 10 'Defoe's England', wrote: 'For Defoe was one of the first who saw the old world through a pair of sharp modern eyes. His report ... occupies the central point of our thought and vision' (pp. 293–4).

Appendix

A selective listing of works connected with the New Sciences in the Defoe/Farewell libraries[1]

Chemistry, physics and mathematics

R. Boyle	*Hydrostatical Paradoxes* (Oxford, 1666).
J. Flamsteed	*Historia Coelestis Britannica* (London, 1725).
R. Hooke	*The Posthumous Works of Robert Hooke; containing his Cutlerian Lectures, and other Discourses read at the Meetings of the Illustrious Royal Society* ... published by R. Waller (London, 1705).
C. Huygens	*Of the Laws of Chance* (London, 1692).
J. Keill	*Introductio ad Veram Physicam ... C. Huygeniis Theoremata de vi Centrifuga & Motu Circulari Demonstrata* (Oxford, 1705).
W. Molyneux	*Sciothericum Telescopicum ... of Adapting a Telescope to an Horizontal Dial* (Dublin, 1686).
I. Newton	*Philosophiae Naturalis Principia Mathematica* (London, 1726).
W. Oughtred	*Trigonometria* (London, 1657).

[1] From *The Libraries of Daniel Defoe and Phillips Farewell, Olive Payne's Sales Catalogue (1731)*, adapted from Heidenreich, ed., 1970.

W. Petty	*The Discourse made before the Royal Society 26 November 1674 concerning the ... Duplicate Proportions* (London, 1674).
H. Plat	*The Jewell House of Art and Nature: containing Divers Rare and Profitable Inventions, together with ... new Experiments in the Art of Husbandry ...* (London, 1653).
J. Wallis	*Opera Mathematica* (Oxford, 1699).
	Mechanica, Sive de Motu, 3 pts. (London, 1669–71).
W. Whiston	*Praelectiones Astronomicae Cantabrigiae* (Cambridge, 1707).
	Praelectiones Physico-Mathematicae Cambridge (London, 1726)
T. Willis	*Opera Omnia* (Amsterdam, 1682).

Botany, gardening and husbandry

R. Austen	*A Treatise of Fruit-Trees* (Oxford, 1657).
W. Blith	*The English Improver Improved, or the Survey of Husbandry Surveyed* (London, 1652).
J. Evelyn	*Kalendarium Hortense; or, the Gardner's Almanac* (London, 1691).
	Acetaria: a Discourse of Sallets (London, 1699).
N. Grew	*The Comparative Anatomy of Trunks, together with an Account of their Vegetation grounded thereupon* (London, 1675).
S. Hartlib	*S. Hartlib's Legacy, or an Enlargement of the Discourse of Husbandry* (London, 1651).
S. Hartlib and R. Weston	*A Discourse of Husbandrie used in Brabant and Flanders* (London, 1650).
	The Reformed Spiritual Husbandman (London, 1652).
J. Laurence	*Gardening Improv'd* containing 1. The Clergyman's Recreation ... 2. The Gentleman's Recreation ... 3. The Fruit-Garden Kalendar ... 4. The Lady's Recreation ... by J. Evelyn (Dublin, 1719).
H. Plat	*The Garden of Eden* (London, 1653).
John Smith	*England's Improvement Reviv'd* (London, 1670).

Geography, topography, travelling and exploring

W. Betagh	*A Voyage round the World. Being an Account of a remarkable Enterprise, begun in the Year 1719, chiefly to cruise on the Spaniards in the great South Ocean* ... (London, 1728).
W. Camden	*Britannia*, ed. Edmund Gibson (London, 1695).
J. Childrey	*Britannia Baconica, or the Natural Rarities of England, Scotland, and Wales* (London, 1661).
D. Denton	*A Brief Description of New York; formerly called New-Netherlands* ... (London, 1670).
A. O. Exquemelin	*The History of the Bucaniers of America*, 2 volume edition containing 4 parts (London 1699).
A. F. Frézier	*A Voyage to the South-Sea* ... (London, 1717).
J. Fryer	*A New Account of East-India and Persia* (London, 1698).
R. Knox	*An Historical Relation of the Island of Ceylon, in the East Indies* ... (London, 1681).
J. Norden	*Speculum Britanniae ... An Historicall and Chorographicall Description of Middlesex and Hartfordshire* (London, 1723).
J. Ogilby	*Britannia ... or, an Illustration of the Kingdom of England and Dominion of Wales* ... (London, 1698).
G. Plattes	*A Discovery of Subterraneall Treasure, viz. of all Manner of Mines and Minerals* ... (London, 1679).
T. Robinson	*New Observations on the Natural History of this World of Matter* (London, 1696).
P. Rycaut	*The Present State of the Ottoman Empire* ... (London, 1670).
L. Wafer	*A New Voyage and Description of the Isthmus of America* (London, 1699).

Industry, crafts, trade and commerce

J. Child	*Discourse about Trade, wherein the Reduction of Interest of Money ... is recommended* (London, 1691).

	Discourse of the Nature, Use, and Advantages of Trade (London, 1694).
	A New Discourse of Trade (London, 1698).
J. Jonston	*Thaumatographia Naturalis* (Amsterdam, 1665).
C. Merrett	*The Mystery of Vintners; some Observations concerning the Ordering of Wine* (London, 1669).
T. Mun	*England's Treasure by Forraign Trade* (London, 1664).
A. Neri and C. Merrett	*The Art of Glass* (London, 1662).
W. Petty	*Further Observations upon the Dublin Bills, or Accompts of the Houses, ... in that City* (London, 1686).
	Five Essays in Political Arithmetick – Observations upon the Cities of London and Rome (London, 1687).
	Two Essays in Political Arithmetick, concerning the People, Housing, Hospitals ... of London and Paris (London, 1687).
	Political Arithmetick, ... concerning the Extent and Value of Lands, People, Buildings (London, 1691)

Religion

R. Boyle	*Some Motives and Incentives to the Love of God* (London, 1659).
	Some Considerations Touching the Style of the Holy Scripture (London, 1661).
	A Free Inquiry into the Vulgarly Receiv'd Notion of Nature (London, 1686).
J. Wilkins	*A Discourse concerning the Gift of Prayer* (London, 1695).

Sundries

J. Aubrey	*Miscellanies, upon the Following Subjects ...* (London, 1721).
J. Dury	*Conscience Eased, or the main Scruple ... against the taking of the Engagement removed* (London, 1651).

J. Dury and S. Hartlib	*A Brief Relation of that which hath been Lately Attempted to Procure Ecclesiasticall Peace among Protestants* (London, 1641).
J. Evelyn	*Numismata. A Discourse of Medals* (London, 1697).
M. Lister	*Sex Exercitationes Medicinales de Quibusdam Morbis Chronicis* (London, 1694).
J. Locke	*Some Thoughts concerning Education* (London, 1705).
J. Ray	*A Collection of English Proverbs* ... (Cambridge, 1678).
W. Wotton	*Reflections upon Ancient and Modern Learning* (London, 1697).

Bibliography

Except for the works by Defoe, the parentheses give the date of the edition used. For the works by Defoe, the parentheses give the date of publication; the date of the edition used, if different, is given thereafter.

Aarsleff, H. (1964), 'Leibniz on Locke on language', *American Philosophical Quarterly*, 1: 165–88.
Adams, P. G. (1962), *Travelers and Travel Liars 1660–1800*, Berkeley, Calif.
Adolph, R. (1968), *The Rise of Modern Prose Style*, Cambridge, Mass.
Aitken, G. A. (1895) (ed.), *Romances and Narratives by Daniel Defoe*, London.
Anderson, F. H. (1948), *The Philosophy of Francis Bacon*, Chicago, Ill.
Andrade, E. D. Da C. (1960), 'Robert Hooke, F.R.S.', *Notes and Records of the Royal Society*, 15: 137–45.
Andrews, J. H. (1959), 'A case of plagiarism in Defoe's *Tour*', *Notes and Queries*, 6: 399.
Aubrey, John (1949), *Brief Lives* ed. O. L. Dick, London.
 (1683–4), *An Idea of Education of Young Gentlemen*, manuscript, Bodleian Library, Oxford.
Backscheider, P. (1985), 'Daniel Defoe as solitary reader', *Princeton University Library Chronicle*, 46: 178–91.
Bacon, Francis (1857–74), *Works* ed. J. Spedding, R. L. Ellis and D. D. Heath, 14 vols., London.
Baine, R. M. (1967), 'Defoe and the angels', *Texas Studies in Literature and Language*, 9: 345–69.
 (1968), *Daniel Defoe and the Supernatural*, Athens, Ga.

Baridon, M. (1984), 'Le style de Defoe et l'épistémologie de la "New Science"', *Trema*, 9: 119–32.
Bastian, F. (1967), 'Defoe's *Tour* and the historian', *History Today*, 17: 845–51.
— (1981), *Defoe's Early Life*, London.
Beattie, J. (1964), *Other Cultures: Aims, Methods, and Achievements in Social Anthropology*, London.
Birch, Thomas (1756–7), *The History of the Royal Society of London for Improving Natural Knowledge*, London.
Blewett, D. (1979), *Defoe's Art of Fiction: 'Robinson Crusoe', 'Moll Flanders', 'Colonel Jack', and 'Roxana'*, Toronto.
Boate, Gerard (1652), *Ireland's Naturall History*, London.
Bonner, W. H. (1934), *Captain William Dampier*, Stanford, Calif.
Boulton, J. T. (1975) (ed.) *Selected Writings of Daniel Defoe*, Cambridge.
Bowen, E. J. and Hartley, H. (1960), 'The Right Reverend John Wilkins, F.R.S.', *Notes and Records of the Royal Society*, 15: 47–56.
Boyle, Robert (1744), *Works* ed. Thomas Birch, 5 vols., London.
Burnet, Thomas (1684), *Sacred Theory of the Earth*, London.
Bush, D. (1962), *English Literature in the Early Seventeenth Century, 1600–1660*, 2nd edn., Oxford.
Calamy, Edmund (1727), *A Continuation of the Account of the Ministers, Lecturers, etc. who were ejected and silenced after the Restoration in 1660 by or before the Act of Uniformity . . .* 2 vols., London.
— (1775), *Nonconformist's Memorial*, 2 vols., London.
Camden, William (1695; facs. repr. 1971), *Britannia* ed. and trans. Edmund Gibson, Newton Abbot.
Christensen, F. (1946), 'John Wilkins and the Royal Society's reform of prose style', *Modern Language Quarterly*, 7: 179–87 and 279–90.
Churchill, John and Awnsham (1704) (eds.) *A Collection of Voyages and Travels . . .* 4 vols., London.
Clauss, S. (1982), 'John Wilkins' *Essay Towards a Real Character*: its place in the seventeenth-century episteme', *Journal of the History of Ideas*, 43: 531–4.
Comenius, John Amos (1668), *Via Lucis*, Amsterdam.
Cornelius, P. (1965), *Languages in Seventeenth- and Early Eighteenth-Century Imaginary Voyages*, Geneva.
Crowther, J. G. (1960), *Founders of British Science*, London.
Dampier, William (1906), *Voyages* ed. John Masefield, 2 vols., London.
Davies, G. (1950), 'Daniel Defoe's *A Tour thro' the Whole Island of Great Britain*', MP, 48: 21–36.
Debus, A. (1970) (ed.) *Science and Education in the Seventeenth Century: The Webster–Ward Debate*, London.
Defoe, Daniel (1697), *An Essay upon Projects*, in Morley 1889.
— (1701), *The True-Born Englishman. A Satyr*, London.

(1702), *The Shortest-Way with the Dissenters: or Proposals for the Establishment of the Church*, London.

(1703), *An Enquiry into the Case of Mr. Asgil's General Translation: Shewing that 'tis not a nearer way to Heaven than the Grave*, London.

(1704), *More Short-Ways with the Dissenters*, London.

(1704), *The Storm: or, a Collection of the most remarkable Casualties and Disasters which happen'd in the Late Dreadful Tempest, both by Sea and Land*, in *The Novels and Miscellaneous Works of Daniel De Foe*, Bohn's Standard Library, vol. 5, London 1911.

(1704–13; facs. repr. 1938), *The Review* ed. A. W. Secord, 22 vols., New York.

(1705), *The Consolidator: or, Memoirs of Sundry Transactions from the World in the Moon*, in Morley 1889.

(1706), *Caledonia, a Poem in Honour of Scotland, and the Scots Nation*, Edinburgh.

(1712), *The Present State of the Parties in Great Britain: Particularly an Enquiry into the State of the Dissenters in England, and the Presbyterians in Scotland ...* London.

(1713), *A General History of Trade*, London.

(1713), *Proposals for Imploying the Poor. In and about the City of London*, London.

(1719), *The Life and Strange Surprising Adventures of Robinson Crusoe, of York, Mariner*, ed. J. D. Crowley, World's Classics paperback, Oxford, 1981.

(1720), *Serious Reflections during the Life and Surprizing Adventures of Robinson Crusoe, with his Vision of the Angelick World*, in Aitken 1895.

(1720), *An Historical Account of the Voyages and Adventures of Sir Walter Raleigh*, London.

(1720–), *Applebee's Original Weekly Journal*, in Lee 1869.

(1721), *Brief Observations on Trade and Manufactures*, London.

(1723), *An Impartial History of the Life and Actions of Peter Alexowitz, the Present Czar of Muscovy*, London.

(1724), *The Great Law of Subordination consider'd; or, the Insolence and Unsufferable Behaviour of Servants in England duly enquir'd into*, London.

(1724–6), *A Tour thro' the Whole Island of Great Britain* ed. G. H. D. Cole, 2 vols., London, 1927; and ed. P. Rogers, Penguin English Library, Harmondsworth, 1971.

('1725' for 1724), *A New Voyage Round the World by a Course never sailed before*, in *The Novels and Miscellaneous Works of Daniel De Foe*, Bohn's British Classics, vol. 6, London 1856.

(1725–7), *A General History of Discoveries and Improvements, in useful Arts, particularly in the great Branches of Commerce, Navigation, and Plantation, in all Parts of the known World*, London.

(1726), *The Complete English Tradesman*, 2nd edn. 1727, London.

(1726), *Mere Nature Delineated: or, a Body without a Soul*, London.

(1726), *A Brief Case of the Distillers, and of the Distilling Trade in England*, London.
(1727), *A Brief Deduction of the Original, Progress, and Immense Greatness of the British Woollen Manufacture*, London.
(1728), *Augusta Triumphans: or, the Way to make London the most flourishing City in the Universe*, London.
(1728), *Atlas Maritimus and Commercialis; or, a General View of the World, so far as it relates to Trade and Navigation*, London.
(1728), *A Plan of the English Commerce*, London.
(1729), *An Humble Proposal to the People of England, for the Encrease of their Trade, and Encouragement of their Manufactures*, London.
(1730), *A Brief State of the Inland or Home Trade of England*, London.
(1731), *Chickens Feed Capons*, London.
(1890), *The Compleat English Gentleman* ed. Karl D. Bülbring, London.
Dijkstra, B. (1987), *Defoe and Economics: The Fortunes of 'Roxana' in the History of Interpretation*, Basingstoke.
Downie, J. A. (1983), 'Defoe, imperialism, and the travel books reconsidered', *The Yearbook of English Studies*, 13: 66–83.
Dryden, John (1882–93), *Works* ed. Sir Walter Scott and G. Saintsbury, 18 vols., Edinburgh.
Earle, P. (1976), *The World of Defoe*, London.
Edwards, P. (1994), *The Story of the Voyage; Sea-Narratives in Eighteenth-Century England*, Cambridge.
Ellis, F. H. (1985), 'Defoe's "Resignâcon" and the limitations of "Mathematical Plainness"', *RES*, 36: 338–54.
Ellis, R. (1906–8), 'Some incidents in the life of Edward Lhuyd', *Honourable Society of Cymmrodorion*, pp. 1–51.
'Espinasse, M. (1956), *Robert Hooke*, London.
 (1958), 'The decline and fall of Restoration Science', *Past and Present*, 14: 71–89.
Evelyn, John (1664), *Sylva, or a Discourse of Forest-Trees, and the Propagation of Timber in His Majesties Dominions*, London.
 (1674), *Navigation and Commerce, their Original and Progress containing a succinct Account of Traffic in general; its Benefits and Improvements*, London.
 (1825), *The Miscellaneous Writings of John Evelyn* ed. William Upcott, London.
 (1959), *The Diary of John Evelyn* ed. E. D. De Beer, London.
Exquemelin, A. O. (Englished 1684–5), *The Bucaniers of America*, London.
Farrington, B. (1973), *Francis Bacon, Philosopher of Industrial Science*, London.
Feingold, M. (1984), *The Mathematician's Apprenticeship: Science, Universities and Society in England 1560–1640*, Cambridge.
Fisch, H. (1952), 'The Puritans and the reform of prose-style', *ELH*, 19: 229–48.

Fisch, H. and Jones H. W. (1951), 'Bacon's influence on Sprat's *History of the Royal Society*', *Modern Language Quarterly*, 12: 399–406.
Fishman, B. (1973), 'Defoe, Herman Moll, and the Geography of South America', *HLQ*, 36: 227–38.
Fitzmaurice, Lord E. (1895), *The Life of Sir William Petty, 1623–1687*, London.
Frank jr., R. G. (1973), 'Science, medicine and the universities of early modern England', *History of Science*, 6: 194–216 and 239–69.
—— (1980), *Harvey and the Oxford Physiologists: A Study of Scientific Ideas*, Berkeley and Los Angeles.
Frantz, R.W. (1968), *The English Traveller and the Movement of Ideas 1660–1732*, New York.
Fraser, R. (1977), *The Language of Adam*, New York.
Frézier, A. (1717), *A Voyage to the South-Sea ... by Monsieur Frezier, Engineer in Ordinary to the French King*, London.
Fulton, J. (1932), 'Robert Boyle and his influence on thought in the seventeenth century', *Isis*, 18: 77–102.
—— (1960), 'The Honourable Robert Boyle, F.R.S.' *Notes and Records of the Royal Society*, 15: 119–35.
—— (1961), *A Bibliography of the Honourable Robert Boyle*, 2nd edn., Oxford.
Furbank, P. N. and W. R. Owens (1988), *The Canonisation of Daniel Defoe*, London.
Gildon, Charles (1719), *The Life and Strange Surprizing Adventures Mr. D——DeF——, of London*, London.
Girdler, Lew (1953), 'Defoe's education at Newington Green Academy', *SP*, 50: 573–91.
Glanvill, Joseph (1668; facs. repr. 1958), *Plus Ultra* ed. J.I. Cope, Gainesville, Fla.
—— (1885), *Scepsis Scientifica* ed. John Owen, London.
Goldsmith, Oliver (1966), *The Collected Works of Oliver Goldsmith* ed. Arthur Friedman, 5 vols., Oxford.
Gosse, E. (1889), *History of Eighteenth Century Literature*, London.
Gunther, R. T. (1920–67) (ed.) *Early Science in Oxford*, 15 vols., Oxford.
Hall, M. B. (1950), 'Boyle as a theoretical scientist', *Isis*, 41: 261–68.
—— (1965), *Robert Boyle on Natural Philosophy*, Bloomington, Ind.
Halley, Edmond (1700), *Magnetic Chart*, London.
Hartley, H. (1960) (ed.) *The Royal Society: Its Origins and Founders*, London.
Hartlib, Samuel (1648), *A Further Discoverie of the Office of Publick Addresse for Accommodations*, London.
—— (1651), *Samuel Hartlib his Legacy: or An Englargement of the Discourse of Husbandry used in Brabant and Flaunders: wherein are bequeathed to the Common-wealth of England more Outlandish and Domestick Experiments and Secrets, in reference to Universall Husbandry*, London.
—— (1652), *The Reformed Spiritual Husbandman*, London.

(1654), *The True and Readie Way to Learne the Latine Tongue*, London.
Healey, G. H. (1969) (ed.) *The Letters of Daniel Defoe*, Oxford.
Heidenreich, H. (1970) (ed.) *The Libraries of Daniel Defoe and Phillips Farewell, Olive Payne's Sales Catalogue (1731)*, Berlin.
Hesse, M. (1964), 'Francis Bacon's philosophy of science', in B. Vickers 1968.
Hill, C. (1965), *Intellectual Origins of the English Revolution*, Oxford.
Hooke, Robert (1665; facs. repr. 1961), *Micrographia: or some Physiological Descriptions of Minute Bodies made by Magnifying Glasses with Observations and Inquiries thereupon*, London.
(1705), *The Posthumous Works of Robert Hooke* ed. Richard Waller, London.
(1935), *The Diary of Robert Hooke, MA., MD., F.R.S. (1672–80)* ed. Henry W. Robinson and Walter Adams, London.
Hornberger, T. (1940), 'Introduction to the Compendium Physicae', *Publications of the Colonial Society of Massachusetts*, 33: xxxi–xl.
Houghton, John (1681–3) (ed.) *A Collection of Letters for the Improvement of Husbandry and Trade*, 2 vols., London.
(1692–1703), *A Collection for Improvement of Husbandry and Trade*, London.
Houghton, W. E. (1941), 'The History of Trades: its relation to seventeenth-century thought as seen in Bacon, Petty, Evelyn and Boyle', *Journal of the History of Ideas*, 2: 33–60.
(1942), 'The English virtuoso in the seventeenth century', *Journal of the History of Ideas*, 3: 51–73 and 190–219.
Howell, A. C. (1946), 'Res et verba: words and things', *ELH*, 13: 131–42.
Howse, D. (1990) (ed.) *Background to Discovery: Pacific Exploration from Dampier to Cook*, Berkeley and Los Angeles, Calif.
Huddlestone, J. (1978), 'Defoe and Charles Morton', *Notes and Queries*, 25: 37–8.
Hunter, J. P. (1966), *The Reluctant Pilgrim: Defoe's Emblematic Method and Quest for Form in 'Robinson Crusoe'*, Baltimore, Md.
Hunter, M. (1975), *John Aubrey and the Realm of Learning*, London.
(1981), *Science and Society in Restoration England*, Cambridge.
Jack, J. (1961), 'A New Voyage round the World: Defoe's *roman à thèse*', *HLQ*, 24: 323–36.
Jones, R. F. (1951), *The Seventeenth Century: Studies in the History of English Thought and Literature from Bacon to Pope*, Stanford, Calif.
(1961), *Ancients and Moderns: A Study of the Rise of the Scientific Movement in Seventeenth-Century England*, St Louis, Mo.
Josselyn, John (1672), *New-Englands Rarities Discovered...* London.
Joyce, James (1964), 'Daniel Defoe' ed. and trans. Joseph Prescott, *Buffalo Studies*, 1: 3–25.
Knowlson, J. (1975), *Universal Language Schemes in England and France 1600–1800*, Toronto.
Knox-Shaw, P. (1987), *The Explorer in English Fiction*, London.

Korshin, P. (1972) (ed.) *Studies in Change and Revolution; Aspects of English Intellectual History, 1640–1800*, Menston.
Lee, W. (1869) (ed.) *Daniel Defoe, His Life, and Recently Discovered Writings: Extending from 1716 to 1729*, 3 vols., London.
Leinster-Mackay, D. P. (1981), *The Educational World of Daniel Defoe*, Victoria, BC.
Levine, J. M. (1972), 'Ancients, moderns and history' in Korshin 1972.
  (1977), *Dr. Woodward's Shield: History, Science, and Satire in Augustan England*, Berkeley and Los Angeles, Calif.
Locke, John (1922), *Educational Writings* ed. J.W. Adamson, 2nd edn., Cambridge.
  1975; repr. (1979), *An Essay Concerning Human Understanding* ed. P. H. Nidditch, Oxford.
McAdoo, H. R. (1965), *The Spirit of Anglicanism: A Survey of Anglican Theological Method in the Seventeenth Century*, London.
Mack, M. and Gregor, I. (1968) (eds.) *Imagined Worlds: Essays on some English Novels and Novelists in Honour of John Butt*, London.
McLachlan, H. (1931), *English Education under the Test Acts*, Manchester.
McVeagh, J. (1981), *Tradeful Merchants, The Portrayal of the Capitalist in Literature*, London.
Maddison, R. E. W. (1961), 'The earliest published writing of Robert Boyle', *Annals of Science*, 17: 165–73.
Masson, I. and Youngson, A. J. (1960), 'Sir William Petty, F.R.S.', *Notes and Records of the Royal Society*, 5: 79–90.
Matthews, A. G. (1934), *Calamy Revised*, Oxford.
Meier, T. K. (1987), *Defoe and the Defence of Commerce*, Victoria, BC.
Merrett, R. J. (1980), *Daniel Defoe's Moral and Rhetorical Ideas*, Victoria, BC.
Milton, John (1973), *Of Education* ed. K. M. Lea, Oxford.
Moll, Herman (1709), *The Compleat Geographer*, London.
Moore, J. R. (1943), *Defoe's Sources for Robert Drury's Journal*, Bloomington, Ind.
  (1958), *Daniel Defoe: Citizen of the Modern World*, Chicago, Ill.
  (1960; rev. 1971), *A Checklist of the Writings of Daniel Defoe*, Hamden, Conn.
Morison, S. E. (1936), *Harvard College in the Seventeenth Century*, 2 pts, Cambridge, Mass.
  (1940), 'Biographical sketch of Charles Morton', *Publications of the Colonial Society of Massachusetts*, 33: vi–xxix.
Morley, H. (1889) (ed.) *The Earlier Life and the Chief Earlier Works of Daniel Defoe*, London.
Morton, Charles (1693), *The Spirit of Man*, Boston, Mass.
  (1727), *Advice to Candidates for the Ministry, under the present discouraging Circumstances* in Calamy 1727, London; written *c.* 1670–80.
  (1940), *The Compendium Physicae* ed. S.E. Morison in *Publications of the Colonial Society of Massachusetts*, 33: 3–237.

Narborough, Sir John (1694), *A Voyage to the South-Sea* ... in *An Account of several late Voyages and Discoveries to the South and North towards the Streights of Magellan* ... London.
Nicolson, M. and Mohler, N. M. (1937), 'The scientific background of Swift's *Voyage to Laputa*', *Annals of Science*, 2: 299–334 and 405–30.
Novak, M. E. (1962; repr. 1976), *Economics and the Fiction of Daniel Defoe*, New York.
 (1963), *Defoe and the Nature of Man*, Oxford.
 (1964), 'Defoe's theory of fiction', *SP*, 61: 650–68.
 (1971), 'Daniel Defoe', *The New Cambridge Bibliography of English Literature* ed. G. Watson, 5 vols., Cambridge, vol. II, cols. 880–917.
 (1983), *Realism, Myth, and History in Defoe's Fiction*, Lincoln, Nebr. and London.
Parker, I. (1914), *Dissenting Academies in England*, Cambridge.
Payne, W. L. (1951) (ed.) *The Best of Defoe's Review*, New York.
 (1961), *Mr. Review*, New York.
Petty, William (1648), *The Advice of W.P. to Mr. Samuel Hartlib for the Advancement of some particular Parts of Learning*, London.
 (1899), *The Economic Writings of Sir William Petty* ed. C. H. Hull, 2 vols., Cambridge.
Philosophical Transactions: giving some account of the present undertakings, studies, and labours of the ingenious in many considerable parts of the world (1665–), London.
Pitt, Moses (1680–3), *The English Atlas*, 4 vols., London.
Plot, Robert (1677), *The Natural History of Oxford-shire*, London.
 (1686), *The Natural History of Stafford-shire*, London.
Pooley, R. (1980), 'Language and loyalty: plain style at the Restoration', *Literature and History*, 6: 2–18.
Power, Henry (1664), *Experimental Philosophy*, London.
Raven, C. E. (1950), *John Ray, Naturalist*, 2nd edn., Cambridge.
Ray, John (1691), *The Wisdom of God manifested in the Works of the Creation*, London.
 (1692; rev. 1693), *Three Physico-Theological Discourses*, London.
Richetti, J. J. (1969), *Popular Fiction before Richardson*, Oxford.
 (1975), *Defoe's Narratives*, Oxford.
Robinson, H. W. (1949), 'An unpublished letter of Dr. Seth Ward relating to the early meetings of the Oxford Philosophical Society', *Notes and Records of the Royal Society*, 7: 68–70.
Rogers, P. (1971), 'Defoe and Virgil: the Georgic element in *A Tour thro' Great Britain*', *English Miscellany*, 22: 93–106.
 (1972 (1)), *Grub Street: Studies in a Subculture*, London.
 (1972 (2)) (ed.) *Defoe: The Critical Heritage*, London and Boston, Mass.
 (1972–3), 'Literary art in Defoe's *Tour*: the rhetoric of growth and decay', *Eighteenth-Century Studies*, 6: 153–85.

(1973), 'Defoe as plagiarist: Camden's Britannia and *A Tour thro' the Whole Island of Great Britain*', *PQ*, 52: 771–4.
(1974 (1)), 'Crusoe's home', *EIC*, 24: 375–90.
(1974 (2)), *The Augustan Vision*, London.
(1974–5), 'Defoe at work: the making of *A Tour thro' Great Britain*, volume 1', *Bulletin of the New York Public Library*, 78: 431–50.
(1975), 'Moll's memory', *English*, 24: 67–72.
(1978) (ed.) *The Eighteenth Century*, London.
(1979), *Robinson Crusoe*, London.
(1980), 'Windsor-Forest, Britannia and river poetry', *SP*, 77: 283–99.
Rogers, Woodes (1718), *A Cruising Voyage Round the World*, London.
Ross, J. F. (1941), *Swift and Defoe: A Study in Relationship*, Berkeley, Calif.
Rousseau, G. S. (1978), 'Science', in Rogers 1978.
Rousseau, Jean-Jacques (1911), *Emile*, trans. B. Foxley, London.
Salmon, V. (1979), *The Study of Language in Seventeenth-Century England*, Amsterdam.
Sawday, J. (1983), 'The mint at Segovia: Digby, Hobbes, Charleton, and the body as a machine in the seventeenth century', *Prose Studies*, 6: 21–35.
Secord, A. W. (1924), *Studies in the Narrative Method of Defoe*, Urbana, Ill.
(1951), 'Defoe in Stoke Newington', *PMLA*, 66: 211–25.
Shapiro, B. J. (1965), *John Wilkins 1614–1672. An Intellectual Biography*, Berkeley and Los Angeles, Calif.
(1983), *Probability and Certainty in Seventeenth-Century England*, Princeton, NJ.
Shinagle, M. (1968), *Daniel Defoe and Middle-Class Gentility*, Cambridge, Mass.
Shipman, J. C. (1962), *William Dampier: Seaman-Scientist*, Lawrence, Kan.
Singer, T. (1989), 'Hieroglyphics, real characters, and the idea of natural language in English seventeenth-century thought', *Journal of the History of Ideas*, 50: 49–70.
Slaughter, M. (1982), *Universal Languages and Scientific Taxonomy in the Seventeenth Century*, Cambridge.
Smith, G. (1985), *The Novel and Society. Defoe to George Eliot*, London.
Smith, J. W. A. (1954) *The Birth of Modern Education*, London.
Snell, G. (1649), *The Right Teaching of Useful Knowledge*, London.
Snow, M. C. (1976), 'The origins of Defoe's first-person narrative technique: an overlooked aspect of the rise of the novel', *The Journal of Narrative Technique*, 6: 175–87.
Sprat, Thomas (1959), *The History of the Royal Society* ed. J. I. Cope and H. W. Jones, St Louis, Mo.
Stamm, R. G. (1936), 'Daniel Defoe: an artist in the Puritan tradition', *PQ*, 15: 225–46.
Starr, G. A. (1965), *Defoe and Spiritual Autobiography*, Princeton, NJ.
(1971), *Defoe and Casuistry*, Princeton, NJ.

(1974), 'Defoe's prose style: 1. The language of interpretation', *MP*, 71: 277–94.
Stimson, D. (1931), 'Dr. Wilkins and the Royal Society', *Journal of Modern History*, 3: 539–63.
(1948), *Scientists and Amateurs: A History of the Royal Society*, New York.
Sutherland, J. (1950), *Defoe*, 2nd edn, London.
(1968), 'The relation of Defoe's fiction to his non-fictional writings', in M. Mack and I. Gregor 1968.
(1971), *Daniel Defoe: A Critical Study*, Cambridge, Mass.
Swift, Jonathan (1972), *Gulliver's Travels* ed. A. Ross, London.
Syfret, R. H. (1947–8), 'The origins of the Royal Society', *Notes and Records of the Royal Society*, 5: 75–137.
Tavor, E. (1987), *Scepticism, Society and the Eighteen-Century Novel*, London.
Thrower, J. W. (1978) (ed.) *The Compleat Plattmaker*, Los Angeles.
Trent, W. P. (1916), *Daniel Defoe: How to Know Him*, Indianapolis, Ind.
Trevelyan, G .M. (1946), *English Social History*, 3rd edn, London.
Turnbull, G. H. (1920), *Samuel Hartlib, A Sketch of his Life and his Relations to J. A. Comenius*, Oxford.
(1947), *Hartlib, Dury and Comenius. Gleanings from Hartlib's Papers*, London.
(1952–3), 'Samuel Hartlib's influence on the early history of the Royal Society', *Notes and Records of the Royal Society*, 10: 101–30.
Vickers, B. (1968) (ed.) *Essential Articles for the Study of Francis Bacon*, Hamden, Conn.
(1985), 'The Royal Society and English prose style: a reassessment', in *Rhetoric and the Pursuit of Truth: Language Change in the Seventeenth and Eighteenth Centuries*, Los Angeles, Calif.
(1987) (ed.) *English Science, Bacon to Newton*, Cambridge.
Vickers, I. (1985), 'A source for Moll Flanders' experience in Virginia', *British Journal for Eighteenth-Century Studies*, 8: 191–4.
(1987), 'The influence of the New Sciences on Defoe', *Literature and History*, 13: 200–18.
Vickery, B. C. (1953), 'The significance of John Wilkins in the history of bibliographical classification', *Libri*, 2: 326–43.
Wafer, L. (1934), *A New Voyage and Description of the Isthmus of America* ed. L. E. Elliott Joyce, Oxford.
Ward, Seth and Wilkins, John (1654), *Vindiciae Academiarum*, Oxford.
Watt, I. (1951), 'Robinson Crusoe as a myth', *EIC*, 1: 95–119.
(1957), 'Defoe as novelist', *The Pelican Guide to English Literature*, vol. 4, ed. B. Ford, Harmondsworth.
(1967), 'Serious reflections on The Rise of the Novel', *Novel: A Forum on Fiction*, 1: 205–18.
(1968) (ed.) *The Augustan Age*, Hamden, Conn.
(1972), *The Rise of the Novel*, London.

Webster, C. (1968–9), 'Henry More and Descartes: some new sources', *British Journal for the History of Science*, 4: 359–77.
  (1970), *Samuel Hartlib and the Advancement of Learning*, Cambridge.
  (1975), *The Great Instauration; Science, Medicine, Reform, 1620–1660*, London.
Webster, John (1654), *Academiarum Examen*, London.
Wesley, Samuel (1704), *Letter from a Country Divine to his Friend in London*, 2nd edn, London.
Wilkins, John (1638), *The Discovery of a World in the Moon*, London.
  (1640), *A Discourse Concerning a New World and another Planet*, London.
  (1641), *Mercury, or the Secret and Swift Messenger*, London.
  (1646), *Ecclesiastes; or, A Discourse concerning the Gift of Preaching, as it falls under the Rules of Art*, London.
  (1648), *Mathematical Magick; or the Wonders that may be performed by Mechanical Geometry*, London.
  (1649), *A Discourse concerning the Beauty of Providence*, London.
  (1651), *Discourse concerning the Gift of Prayer*, London.
  (1668), *Essay Towards a Real Character and a Philosophical Language*, London.
  (1802; facs. repr. 1970), *The Mathematical and Philosophical Works of the Right Rev. John Wilkins*, London.
Willey, B. (1940; repr. 1980) *The Eighteenth Century Background: Studies on the Idea of Nature in the Thought of the Period*, London.
Williams, G. (1990), 'The achievement of the English voyages 1650–1800', in Howse 1990.
Woodward, John (1695), *An Essay toward a Natural History of the Earth and Terrestrial Bodies* ... London.
  (1696), *Brief Instructions for Making Observations in all Parts of the World: as also for Collecting, Preserving and Sending over Natural Things* ... London.

Index

Page numbers in italic refer to illustrations

Aarsleff, H., 124
Addison, Joseph, 96
Adolph, R., 124
ancients and moderns, 16, 74–5
Anderson, F. H., 124
Andrade, E. D. Da C., 27
Andrews, J. H., 161, 164
antiquity, natural history of, 172
Aubrey, John, 59, 67, 132, 161, 172

Backscheider, P., 4, 15n.6
Bacon, Francis, 9–16, 40–2, 43–4, 45–6, 59, 60n.4, 67, 69, 70–1, 73, 74, 75, 82, 83–4, 85–6, 87, 101, 110, 116, 125, 131, 133, 151–2, 153, 175–6
Baridon, M., 4, 124
Bastian, F., 32, 101, 161
Beattie, J., 121
Beaumont, John, 165
Bernard, Edward, 66
Birch, Thomas, 85, 159
Bonner, W. H., 141
Boulton, J. T., 124
Bowen, E. J., 29
Boyle, Robert, 15, 18, 20, 22, 23, 25–6, 31, 40n.10, 42, 44, 67, 68, 70, 74, 75, 76–7, 79, 83, 84, 85, 86, 94, 96, 103–4, 106, 108, 110, 112, 115, 119, 126, 132, 133, 145, 147, 152, 157
Burnet, Thomas, 92
Bush, D., 48

Calamy, Edmund, 32
Camden, William
 Britannia, 68, 159, 160–8, 170
Cato the Elder, 43, 50, 125; *see also res* and *verba*; words and things
Churchill, John and Awnsham, 137–8, 143, 148
Clauss, S., 47
Comenius (or Komensky), John Amos, 18, 20, 21
Cornelius, P., 47
Cox, Daniel, 35
Crowther, J. G., 29

Dampier, Captain William, 134, 135–6, 136–7, 138, 141, 142, 148, 170
Davies, G., 161
Debus, A., 33n.4
Defoe, Daniel
 as traveller-scientist by land, 154–60
 as traveller-scientist by sea, 138–50
 consistency of ideas, 55, 80, 103, 117–18, 148–9, 174–6
 education, 16–7, 55–65, 73, 120–2, 125
 plain prose style, 51, 58, 122–3; *see also* plain prose; words and things
 problem of attribution, 5
 promotion of useful or 'real' knowledge, 56–9, 61–4, 78, 79–80, 96, 97–8, 102–5, 169–76
 reform, economic and social, 73, 75, 78–80, 96–8, 149–50, 169–71, 173–4

Defoe, Daniel (*cont.*)
  savages, Defoe's attitude to, 120–2
  trade and colonisation, 73, 94–6, 141–2, 147–9, 173–5
  use of evidence, 26–7, 65–9, 70–2, 76–8, 101–2, 110, 142–3, 144–8, 149, 154–5, 158, 159–60, 160–9
  WORKS
    Applebee's Journal, 58
    Atlas Maritimus, 24, 175
    Augusta Triumphans, 63–4
    Brief Case of the Distillers, 95
    Brief Deduction of the Original ... Woollen Manufacture, 174
    Brief State of the Inland and Home Trade, 89, 95, 98, 103
    Caledonia, 27, 73, 173, 174
    Captain Singleton, 132, 139, 142
    Chicken Feed Capons, 121
    Complete English Gentleman, 16, 17, 26, 56, 57, 58, 60, 62, 63, 75–6, 77, 80, 101, 103, 122, 131, 138, 174, 175
    Complete English Tradesman, 51, 56, 122–3
    Consolidator, 26, 31, 70–3, 77, 80
    Crusoe I, 4, 27, 30, 99–131
    Crusoe II, 4, 132
    Crusoe III, 51, 101–2, 115, 116, 117, 118, 122, 173, 174
    Enquiry into the Case of Mr Asgil's General Translations, 69
    Essay upon Projects, 24, 31, 60, 62–3, 76, 80, 120
    General History of Discoveries and Improvements, 16, 26, 63, 73–80, 81, 95, 96, 98, 101, 103, 116, 141, 147, 148, 169–70, 171, 174
    General History of Trade, 22, 73, 81, 88–98, *90*, 103, 104, 105–6, 148, 169, 171–2, 174; *see also* Trades, Histories of
    Great Law of Subordination Consider'd, 58, 160
    Historical Account of the Voyages and Adventures of Sir Walter Raleigh, 4, 103, 121, 141
    Humble Proposal, 98, 103, 173–4
    Impartial History of ... Peter Alexowitz, 175
    Mere Nature Delineated, 60, 62, 120, 121
    Moll Flanders, 98
    More Short-ways with the Dissenters, 38–9
    New Voyage Round the World, 30, 40, 132, 138–50, 174
    Plan of the English Commerce, 24, 98, 103, 142, 148
    Present State of the Parties, 36–7, 56, 58, 120
    Proposals for Imploying the Poor, 64
    Review, 24, 60, 63, 94, 106, 125, 141, 155, 173, 175
  Roxana, 98
  Storm, 26, 65, 66–9, 73, 77, 80, 117–18, 122
  Tour, 4, 22, 24, 40, 58, 68, 91, 103, 132, 142, 145, 149, 151–2, 154–76
  True-Born Englishman, 60
  (for the sales catalogue of the Defoe/ Farewell libraries see H. Heidenreich)
Dissenting Academies, 22, 31, 32, 36–8, 56–8, 61–2, 125
Down Survey, 24
Downie, J. A., 120
Dryden, John, 131
Dury, John, 18, 20, 21, 37, 59, 64, 112

Earle, P., 143
Edwards, P., 133
Ellis, F. H., 124
'Espinasse, M., 27
Evelyn, John, 18, 20, 23, 28, 35, 83, 85, 87–8, 91, 92–3, 110, 132, 160, 167
Exquemelin (or Esquemeling), A. O., 134, 143, 148

Feingold, M., 33
Fisch, H., 46, 120
Fishman, B., 142, 143
Fitzmaurice, Lord E., 23
fossils, theories of, 164–5
Frank jr., R. G., 33, 34
Frantz, R. W., 133
Fraser, R., 47
Frézier, A., 134, 148
Fulton, J., 25
Furbank, P. N., 5

Gibson, Edmund, 68, 160–8; *see also* Camden, *Britannia*
Gildon, Charles, 115
Gipseys (Vispeys), 166
Girdler, Lew, 32, 39
Glanvill, Joseph, 12
Gloucester: 'whispering place' at, 39–40, 159
Goddard, J., 23, 34
Godwin, F., 71
Goldsmith, Oliver, 74–5
Gosse, E., 48
Gresham College, 34, 64
Gunther, R. T., 27

Hall, M. B., 26,
Halley, E., 136, 139, *140*
Harley, Robert, Earl of Oxford, 141
Hartlib, Samuel, 18, 20–2, 23, 25, 37–8, 58, 64, 83, 97, 153, 154
Harvard College, New England, 32–3
Harvey, W., 39, 67

Index

Healey, G. H., 125, 131, 141
Heidenreich, H., 5, 19, 22, 24, 27, 29, 31, 49
Hesse, M., 11
Hill, C., 9, 38, 48, 64, 124
History of the Royal Society, 9, 27, 30, 45–6,
    50–1, 65–6, 85, 86, 87, 103, 108, 112,
    113, 123, 127, 133
Hooke, Robert, 18, 25, 27–9, 31, 38, 40, 44,
    45, 46, 59, 70, 84, 85, 90, 94, 96, 103,
    105, 108, *109*, 114, 136, 165, 169,
    172
Hornberger, T., 34, 40
Houghton, John, 87–8, 91
Houghton, W. E., 19, 82
Howell, A. C., 43, 59
Huddlestone, J., 50
Hunter, J. P., 108, 113, 120
Hunter, M., 9, 36, 59, 67, 82, 86, 87

Invisible College, 18

Jack, J., 143
Jones, H. W., 46
Jones, R. F., 9, 11–12, 14, 16, 38, 43
Joyce, James, 131

Knowlson, J., 47
Korshin, P., 16

Laughton, J. K., 136
Lee, W., 32
Leeuwenhoek, Anthony van, 67–8
Leinster-Mackay, D. P., 56
Levine, J. M., 16, 165
Lhwyd, Edward, 66, 132, 136, 160–1, 165,
    168
Lister, Dr Martin, 39, 66
Locke, John, 30n.22, 48, 59, 60–1n.4, 62, 70,
    76, 108, 123, 124

McAdoo, H. R., 120n.15
McLachlan, H., 32, 36
McVeagh, J., 4
Masson, I., 23
Matthews, A. G., 32
Merrett, R. J., 69
Milton, John, 20, 59, 60
Mohler, N. M., 70, 72
Moll, Herman, 142
Moore, J. R., 5, 9, 101, 118, 155
Morison, S. E., 33, 38
Morland, Samuel, 71
Morton, Charles, 22, 26, 31, 32–51, 59, 61,
    62, 70, 71, 79, 108, 113, 125, 131,
    149, 159
WORKS
    Advice to Candidates, 48–51, 59

Compendium Physicae, 26, 38–42, 57, 69,
    70, 71, 168
Spirit of Man, 37, 62

Narborough, Sir John, 134, 143–8
natural theology, 14, 15, 27, 69, 93–4,
    112–20; *see also* providence tradition
Nicolson, M., 70, 72
Novak, M. E., 60–1n.4, 123

Ovalle, Alonso de, 143
Owens, W. R., 5
Oxford, scientific education at, 33–5

Parker, I., 36, 38, 51
petrification, theories of, 167
Petty, Sir William, 18, 20, 22–5, 28, 31, 33,
    34, 37, 39, 46, 59, 64, 78, 83–4, 85,
    110, 129
Philosophical Society of Oxford, 22, 23, 24,
    31, 34, 78
Philosophical Transactions, 35, 39, 40, 65, 67,
    107, *109*, 133, 136, 143, 144, 152,
    157, 168
Pitt, Moses, 152
plain prose, New Sciences' demand for, 21,
    30, 42–51, 55, 59–61, 122–31, 137
Plot, Robert, 132, 151–2, 160, 161, 165,
    172
Pope, W., 34
Power, Henry, 40
providence tradition, 27, 92, 93–4, 113–15,
    119, 173–4; *see also* natural theology
present state, enquiry into, 154, 156, 157,
    172

Ray, John, 15, 66, 94, 114–15, 117, 132, 136,
    161, 164, 165, 166
res and *verba*, 43, 50, 59; *see also* Cato the Elder;
    words and things
Ringrose, Basil, 143
Rogers, P., 118, 128, 130, 142, 155, 161, 164,
    170
Rogers, Woodes, 134, 137, 138, 141, 142,
    148
Rooke, L., 31, 34, 133
Royal Society of London, 19, 22, 23, 24, 25,
    29, 34, 63, 64, 65, 70, 73, 85–6,
    102–3, 106–8, 110, 111–13, 123–4,
    135, 175
    'Directions for Seamen', 65–6, 106–7,
    132–8, 144, 145–6, 148, 151, 152
    'Directions for travellers by land', 133,
    152–4, 158

Salmon, V., 47
Sawday, J., 96

Secord, A. W., 118, 141, 143
Shapiro, B. J., 4, 29, 133
Singer, T., 60–1n.4
Slaughter, M., 47
Smith, J. W. A., 36, 38
Snell, G., 21
Snow, M. C., 124
Solomon (or Salomon), king, 15, 17, 60, 101, 116, 174–5
  Solomon's fool, 102, 116
  Solomon's House, 12, 70–1
'speaking trumpet', 35, 70–1
Spectator, 96
Stamford Oath, 32
Stamm, R. G., 120
Starr, G. A., 108, 113, 120, 124
Stimson, D., 29, 30
Stukeley, William, 172
Sutherland, J., 101
Swift, Jonathan, 16, 48, 72–3
Syfret, R. H., 18, 20, 21, 64

Teague, Baden, 41n.11
Thoresby, Ralph, 68, 132, 161
Thrower, J. W., 136
Trades, History of
  Bacon: 12, 41, 82–3, 103, 105–6
  Boyle: 26, 83, 84, 85, 86, 103, 110
  Defoe: 81–2, 88–98, *90*
  Evelyn: 83, 85, 93, 110
  Hartlib: 20–1, 83
  Houghton: 87–8, 91
  Merrett: 85
  Petty: 23, 83–4, 85, 94
  Wilkins: 30, 85, 110
Trevelyan, G. M., 175–6
Turnbull, G. H., 18, 20, 64

Universal (or Philosophical) Language Schemes, 30, 46–8

Vickers, B., 14, 25, 43, 46
Vickery, B. C., 47

Wadham College, Oxford, 23, 31, 32, 34, 70–1
Wafer, L., 134, 143, 148
Waller, Richard, 29
Wallis, John, 23, 31, 34, 40, 41, 66
Ward, Seth, 20, 31, 33, 34, 78
Watt, I., 123
Weather, History of
  Bacon: 41
  Boyle: 41, 108
  Defoe: 108, 149–50
  Hooke: 30, 70, 108, *109*
  Locke: 30, 70, 108
  Morton: 41, 70, 108, 149
  Wilkins: 30, 70, 108
  Wren: 30, 70, 108
Webster, C., 9, 19, 20, 64, 82, 110, 133
Webster, John, 33
Wesley, Samuel, 32, 38, 71
Wilkins, John, 15, 18, 20, 23, 25, 28, 29–31, 33, 34, 37, 43, 44, 45–6, 48–50, 70–2, 85, 94, 108, 110, 113, 119, 125
Williams, G., 139
Willey, B., 94
Woodward, John, 132, 134, 136, 152–4, 158, 161, 165, 166, 172
words and things, 43, 48, 50, 58–60, 74, 83, 98, 125, 131; *see also res* and *verba*
Wren, Sir Christopher, 20, 25, 28, 29, 30, 31, 34, 35, 59, 70, 108

Youngson, A. J., 23

CAMBRIDGE STUDIES IN EIGHTEENTH-CENTURY
ENGLISH LITERATURE AND THOUGHT

General Editors

Professor HOWARD ERSKINE-HILL LITT.D., FBA, *Pembroke College, Cambridge*
Professor JOHN RICHETTI, *University of Pennsylvania*

1 *The Transformation of* The Decline and Fall of the Roman Empire
by David Womersley
2 *Women's Place in Pope's World*
by Valerie Rumbold
3 *Sterne's Fiction and the Double Principle*
by Jonathan Lamb
4 *Warrior Women and Popular Balladry, 1650–1850*
by Dianne Dugaw
5 *The Body in Swift and Defoe*
by Carol Flynn
6 *The Rhetoric of Berkeley's Philosophy*
by Peter Walmsley
7 *Space and the Eighteenth-Century English Novel*
by Simon Varey
8 *Reason, Grace, and Sentiment*
A Study of the Language of Religion and Ethics in England, 1660–1780
by Isabel Rivers
9 *Defoe's Politics: Parliament, Power, Kingship and* Robinson Crusoe
by Manuel Schonhorn
10 *Sentimental Comedy: Theory & Practice*
by Frank Ellis
11 *Arguments of Augustan Wit*
by John Sitter
12 *Robert South (1634–1716): An Introduction to his Life and Sermons*
by Gerard Reedy, SJ
13 *Richardson's* Clarissa *and the Eighteenth-Century Reader*
by Tom Keymer
14 *Eighteenth-Century Sensibility and the Novel*
The Senses in Social Context
by Ann Jessie Van Sant
15 *Family and the Law in Eighteenth-Century Fiction*
The Public Conscience in the Private Sphere
by John P. Zomchick

16 *Crime and Defoe: A New Kind of Writing*
by Lincoln B. Faller
17 *Literary Transmission and Authority: Dryden and Other Writers*
edited by Earl Minor and Jennifer Brady
18 *Plots and Counterplots*
Sexual Politics and the Body Politic in English Literature, 1660–1730
by Richard Braverman
19 *The Eighteenth-Century Hymn in England*
by Donald Davie
20 *Swift's Politics: A Study in Disaffection*
by Ian Higgins
21 *Writings and the Rise of Finance*
Capital Satires of the Early Eighteenth-Century
by Colin Nicholson
22 *Locke, Literary Criticism, and Philosophy*
by William Walker
23 *Poetry and Jacobite Politics in Eighteenth-Century Britain and Ireland*
by Murray G. H. Pittock
24 *The Story of the Voyage in Eighteenth-Century England*
by Philip Edwards
25 *Edmond Malone: A Literary Biography*
by Peter Martin
26 *Swift's Parody*
by Robert Phiddian
27 *Rural Life in Eighteenth-Century English Poetry*
by John Goodridge
28 *The English Fable: Aesop and Literary Culture, 1651–1740*
by Jayne Elizabeth Lewis
29 *Mania and Literary Style*
The Rhetoric of Enthusiasm from the Ranters to Christopher Smart
by Clement Hawes
30 *Landscape, Liberty and Authority*
Poetry, Criticism and Politics from Thomson to Wordsworth
by Tim Fulford
31 *Philosophical Dialogue in the British Enlightenment*
Theology, Aesthetics, and the Novel
by Michael B. Prince
32 *Defoe and the New Sciences*
by Ilse Vickers